Fluid Mechanics of Plankton

Fluid Mechanics of Plankton

Editors

Hidekatsu Yamazaki
J. Rudi Strickler

MDPI • Basel • Beijing • Wuhan • Barcelona • Belgrade • Manchester • Tokyo • Cluj • Tianjin

Editors
Hidekatsu Yamazaki
Shanghai Ocean University
China

J. Rudi Strickler
University of
Wisconsin-Milwaukee
USA

Editorial Office
MDPI
St. Alban-Anlage 66
4052 Basel, Switzerland

This is a reprint of articles published online by the open access publisher MDPI (available at: https://www.mdpi.com/journal/fluids/special_issues/fluid_mechanics_plankton). The responsibility for the book's title and preface lies with Hidekatsu Yamazaki, who compiled this selection.

For citation purposes, cite each article independently as indicated on the article page online and as indicated below:

LastName, A.A.; LastName, B.B.; LastName, C.C. Article Title. *Journal Name* **Year**, *Volume Number*, Page Range.

ISBN 978-3-0365-0836-8 (Hbk)
ISBN 978-3-0365-0837-5 (PDF)

Cover image courtesy of J. Rudi Strickler.

© 2021 by the authors. Articles in this book are Open Access and distributed under the Creative Commons Attribution (CC BY) license, which allows users to download, copy and build upon published articles, as long as the author and publisher are properly credited, which ensures maximum dissemination and a wider impact of our publications.
The book as a whole is distributed by MDPI under the terms and conditions of the Creative Commons license CC BY-NC-ND.

Contents

About the Editors vii

Preface to "Fluid Mechanics of Plankton" ix

Hidekatsu Yamazaki and J. Rudi Strickler
Fluid Mechanics of Plankton
Reprinted from: *Fluids* **2021**, *6*, 56, doi:10.3390/fluids6020056 1

Mamoru Tanaka
Changes in Vertical Distribution of Zooplankton under Wind-Induced Turbulence: A 36-Year Record
Reprinted from: *Fluids* **2019**, *4*, 195, doi:10.3390/fluids4040195 5

Kacie T. M. Niimoto, Kyleigh J. Kuball, Lauren N. Block, Petra H. Lenz and Daisuke Takagi
Rotational Maneuvers of Copepod Nauplii at Low Reynolds Number
Reprinted from: *Fluids* **2020**, *5*, 78, doi:10.3390/fluids5020078 17

John O. Dabiri, Sean P. Colin, Brad J. Gemmell, Kelsey N. Lucas, Megan C. Leftwich. and John H. Costello
Jellyfish and Fish Solve the Challenges of Turning Dynamics Similarly to Achieve High Maneuverability
Reprinted from: *Fluids* **2020**, *5*, 106, doi:10.3390/fluids5030106 31

Houshuo Jiang
An Elastic Collision Model for Impulsive Jumping by Small Planktonic Organisms
Reprinted from: *Fluids* **2020**, *5*, 154, doi:10.3390/fluids5030154 45

Leonid Svetlichny, Poul S. Larsen and Thomas Kiørboe
Kinematic and Dynamic Scaling of Copepod Swimming
Reprinted from: *Fluids* **2020**, *5*, 68, doi:10.3390/fluids5020068 61

Caroline H. Suwaki, Leandro T. De-La-Cruz and Rubens M. Lopes
Impacts of Microplastics on the Swimming Behavior of the Copepod *Temora turbinata* (Dana, 1849)
Reprinted from: *Fluids* **2020**, *5*, 103, doi:10.3390/fluids5030103 89

Erik Selander, Sam T. Fredriksson and Lars Arneborg
Chemical Signaling in the Turbulent Ocean—Hide and Seek at the Kolmogorov Scale
Reprinted from: *Fluids* **2020**, *5*, 54, doi:10.3390/fluids5020054 99

Iman Borazjani
Numerical Simulations of Flow around Copepods: Challenges and Future Directions
Reprinted from: *Fluids* **2020**, *5*, 52, doi:10.3390/fluids5020052 111

Mathilde Schapira and Laurent Seuront
Nutrient Patchiness, Phytoplankton Surge-Uptake, and Turbulent History: A Theoretical Approach and Its Experimental Validation
Reprinted from: *Fluids* **2020**, *5*, 80, doi:10.3390/fluids5020080 123

Zachary Wagner, John H. Costello and Sean P. Colin
Fluid and Predator-Prey Interactions of Scyphomedusae Fed Calanoid Copepods
Reprinted from: *Fluids* **2020**, *5*, 60, doi:10.3390/fluids5020060 141

Hans L. Pécseli, Jan K. Trulsen, Jan Erik Stiansen and Svein Sundby
Feeding of Plankton in a Turbulent Environment:A Comparison of Analytical and Observational ResultsCovering Also Strong Turbulence
Reprinted from: *Fluids* **2020**, 5, 37, doi:10.3390/fluids5010037 . . . . **153**

About the Editors

Hidekatsu Yamazaki is distinguished professor at College of Marine Ecology and Environment, Shanghai Ocean University and emeritus professor of Department of Ocean Sciences, Tokyo University of Marine Science and Technology. He received Ph.D. in ocean engineering from Texas A&M University in 1984. He worked at Department of Oceanography, Naval Postgraduate School (NPS) and switched his expertise from ocean engineering to oceanography. His interest in oceanic microstructures, particularly turbulence and plankton, started at NPS. He also worked at Chesapeake Bay Institute, Johns Hopkins University and School of Earth and Ocean Science, University of Victoria before he returned to Japan in 1993. His research spans from oceanic microstructures to fisheries ground environments as well as various biophysical coupling problems.

J. Rudi Strickler (officially Johann Rudolf Strickler), born and raised in Switzerland. 1969 Dr.sc.nat. at the Swiss Federal Institute of Technology in Zurich (ETH-Z). Former positions: Principal Research Scientist, Australian Institute of Marine Sciences. Faculty positions at Johns Hopkins University, Yale University, University of Ottawa, University of Southern California, Boston University. Currently Shaw Distinguished Professor at the University of Wisconsin-Milwaukee and Adjunct Professor at the University of Texas at Austin. Research on the biology-physics intersection at the temporal and spatial scales of the zooplankton.

Preface to "Fluid Mechanics of Plankton"

This book focuses on the question about moving "with a certain degree of independence from the driving forces." The perception that some animals just drift with the water flow is a scale-dependent question. Looking at the planet Earth from the star Sirius would not give the observer the impression that we human entities move around. Our contributions look at different scales to comprehend the importance of plankton to the life in the oceans.

Hidekatsu Yamazaki, J. Rudi Strickler
Editors

Editorial

Fluid Mechanics of Plankton

Hidekatsu Yamazaki [1,2,3,*] and J. Rudi Strickler [4,5,*]

[1] College of Marine Ecology and Environment, Shanghai Ocean University, Shanghai 201306, China
[2] Alpha Hydraulic Engineering Consultants Co. Ltd., Chuoh-ku, Tokyo 104-0045, Japan
[3] Department of Ocean Sciences, Tokyo University of Marine Science and Technology, Minato-ku, Tokyo 108-8447, Japan
[4] Department of Biological Sciences, University of Wisconsin-Milwaukee, Milwaukee, WI 53211, USA
[5] Marine Science Institute, University of Texas at Austin, Port Aransas, TX 78712, USA
* Correspondence: hide@kaiyodai.ac.jp (H.Y.); jrs@uwm.edu (J.R.S.)

Received: 21 January 2021; Accepted: 22 January 2021; Published: 27 January 2021

These first lines of Hensen's article (Figure 1) in the "Fünfter Bericht" (1887) translate as follows.

"The material called "Auftrieb" has been investigated by zoologists and botanists since the groundbreaking contributions by Johannes Müller. It has been investigated and collected many times with fine, permeable nets. This material is—besides Müller's interest in its systematics and anatomy—without any doubt of great importance to the metabolism of the seas.

This contribution tries to get a closer look at this metabolism. It turns out that the name "Auftrieb" is not sufficiently comprehensive and descriptive, therefore, I have preferred to name this material "Halyplankton" (1). However, since we are only talking about the oceans here, the shorter term "Plankton" will be sufficient. It is defined as "everything that floats in water, regardless of whether it is high or low in it, and whether it is dead or alive."

A limitation of the expression to include only certain forms as plankton would not encompass the many embryonic forms that no longer occur in the plankton in their developed states. The decisive factor is whether the animals drift with the flow of the water, or whether they move with a certain degree of independence from the driving forces? Fish, therefore, belong only in the form of eggs and brood to plankton, but not as adult animals; whereas the copepods, though lively swimming, are carried away with the water flow, and must therefore be counted as part of the plankton . . . "

The change of words, from "Auftrieb" to "Plankton", was a big step in the perception of life in the oceans. "Auftrieb", the German word for buoyancy, defined anything floating in the water due to buoyancy. Scientists, like Johannes Müller mentioned in the article, were interested in the different animals floating within the material. They researched physiological and morphological questions, wondered about mating mechanics, but did not put the different species in context with each other. With naming this community of floating biology "Plankton" a first step toward recognizing an ecosystem was made.

Hensen continued these starting lines with a short discussion about how deep light might penetrate the upper water layer, and therefore, living entities may use it for living. He concluded that life in the oceans, and especially to produce fish, depends on plankton as the source of food (Figure 2).

With this step, Hensen accepted plankton as a lower level of the food pyramid and became the "grandfather of biological oceanography".

Life of the earth appeared roughly 3.8 billion years ago and spent nearly 3 billion years as single-cell organisms. Due to the limitation of molecular diffusion, a single cell can be no more than 1 mm scale, in which the viscosity dominates (the Kolmogorov scale). Under this condition, life stayed another nearly two billion years before multicell organisms emerged. At the present date, most phytoplankton cells are below the Kolmogorov scale. Even a large faction of zooplankton is also not free from the viscosity of water. It would be unwise to assume that all microscale organisms do not pay attention to the immediate surrounding fluid motions. In fact, many microscale organisms swim more than 10 body-lengths per second, whereas most large-scale organisms swim an order of one body-length per second. Clearly, microorganisms have the ability to manipulate the properties of water. In order to maintain the population, phytoplankton require sunlight and nutrients; zooplankton pay attention to feeding, mating and escaping from predators.

Das mit dem Namen »Auftrieb« bezeichnete Material, welches von Zoologen und Botanikern, seit den bahnbrechenden Arbeiten von JOHANNES MÜLLER, so vielfach untersucht worden ist und mit feinem durchlässigen Netz gefischt wird, hat, neben seinem Interesse für Systematik und Anatomie, ohne Zweifel für den ganzen Stoffwechsel des Meeres eine grosse Bedeutung.

Diese Arbeit versucht es, dem bezüglichen Stoffwechsel näher zu treten. Es zeigt sich, dass der Name: Auftrieb nicht genügend umfassend und bequem ist, so dass ich es vorgezogen habe dies Material mit dem Namen Halyplankton [1]) zu belegen; da hier jedoch nur vom Meere die Rede ist genügt der kürzere Ausdruck Plankton. Ich verstehe darunter Alles was im Wasser treibt, einerlei ob hoch oder tief, ob todt oder lebendig.

Eine bestimmte Begrenzung des Ausdrucks auf bestimmte Formen lässt sich nicht gewinnen, denn es gehören zum Plankton viele Embryonal-Formen, die im entwickelten Zustand ihm Plankton nicht mehr vorkommen. Das Entscheidende ist der Umstand, ob die Thiere willenlos mit dem Wasser treiben oder ob sie einen gewissen Grad der Selbständigkeit dieser Triebkraft gegenüber bewahren? Die Fische gehören daher höchstens in der Form von Eiern und Brut zum Plankton, aber nicht als erwachsene Thiere; die *Copepoden*, obgleich lebhaft schwimmend, werden doch willenlos mit dem Wasser fortgerissen, und müssen daher zum Plankton gerechnet werden. Auf eine ganz genaue Abgrenzung des Begriffes kommt es durchaus nicht an; es knüpft sich dagegen,

[1]) Der Ausdruck ist nach einem Vorschlag meines Kollegen Professor FÖRSTER aus ἡ ἅλς und πλανάω gebildet. Der Ausdruck pelagisch bezeichnet nur das ganz allgemein zum Meere Gehörige.

Figure 1. First lines of Hensen, Victor. 1887. "Ueber die Bestimmung des Plankton's oder des im Meere treibenden Materials an Pflanzen und Thieren". *Bericht der Kommission zur wissenschaftlichen Untersuchungen der deutschen Meere, in Kiel* 5: 1–107, 6 pls.

erkläre mir dies Verhalten daraus, dass ein in fauliger Zersetzung begriffenes Material einer stärkeren Verdauungskraft bedarf, als diejenige ist, welche die Organisation der niederen Thiere hervorzubringen vermag.

Das Plankton kann im Allgemeinen wohl überall im Meere wachsen, es gewährt lebendige Nahrung, und scheint deshalb eine vorzügliche Quelle der Nahrungsstoffe zu sein.

Dieser Stoffquelle nachzugehen ist die Aufgabe, zu welcher ich im Verlauf meiner Untersuchungen geführt worden bin. Schon als ich 1867 im preussischen Landtag für die Herbeiziehung wissenschaftlicher Untersucher zur Erforschung des Meeres im Interesse der Fischerei eintrat, schwebte mir der Gedanke vor, dass man nur dann richtige Massnahmen im Interesse der Fischerei werde aufzufinden vermögen, wenn man in der Lage sei, sich ein Urtheil über die Ertragsfähigkeit des Meeres zu bilden. Als Mitglied der bezüglichen wissenschaftlichen

Figure 2. Excerpt from Hensen (1887). The small paragraph translates as: "Plankton can in principle grow all over in the oceans. It constitutes live food and therefore is a great source of nutrition".

Our contribution here in this Special Issue focuses on some of the few words at the end of the text in Figure 1. It is the question about moving "with a certain degree of independence from the driving forces." The perception that some animals just drift with the water flow is a scale-dependent question. Looking at the planet Earth from the star Sirius would not give the observer the impression that we human entities move around. Our contributions look at different scales to comprehend the importance of plankton to the life in the oceans as Hensen years ago was already contemplating.

Making use of a long observational data set, Tanaka [1] shows evidence of turbulence avoidance, and found that ambush feeders showed statistically significant changes in response to turbulence, whereas suspension feeders did not. Niimoto et al. [2] demonstrated the physical mechanisms of rotation for copepod nauplii about three principal axes of the body: yaw, roll, and pitch. Based on the results gained in experiments, they suggest the development of microscopic robots. Dabiri et al. [3] studied turning mechanisms of aquatic animals, e.g., jellyfish, and zebrafish. Turning requires torque while minimizing the resistance to the moment of inertia. These two are opposing mechanisms. Their results are based on laboratory experiments and show how aquatic animals balance these mechanisms. Jiang [4] proposes a theoretical fluid mechanics model to estimate propulsion efficiencies of several planktonic species. For example, a tailed ciliate shows a high efficiency (~0.9). Copepods also show an unexpectedly high efficiency (>0.95), whereas in squid it is 0.44 and in small medusae 0.38.

Svetlichny et al. [5] provides scaling laws for swimming modes of calanoid copepods. Cruise swimming and short-lasting jumps are scaled with prosome length to a power between 2 and 3. The cost of transportation was higher for jumping than cruise swimming by a factor of 7 for large copepods but only a factor of 3 for small ones. These facts explain why small copepods can afford to more often jump than large copepods that are cruising swimmers. Suwaki et al. [6] studied the potential impacts of microplastics on zooplankton behavior. Microplastics are a recent hot topic because of human's interference with oceanic life. They found that the swimming behavior of calanoid copepod *Temora turbinate* is affected by microplastics. Selander et al. [7] discussed chemical cues that are used for resource acquisition, mate finding and assessing predation risk. They investigated how turbulence affects the distribution of chemical properties at a micro-scale using a numerical simulation model (DNS). They found that the chemical trail can be found under moderate turbulence conditions, but, when the rate of turbulent kinetic energy dissipation exceeds 10^{-7} (W·kg^{-1}), the trails are shortened drastically.

Borazjani [8] reviewed numerical methods used to study (1) the force and flow generated by different part of body; (2) the relation between the small-scale flow around the body and the large-scale flow; and (3) flow and energetics. The author also discusses future prospects of numerical model developments. Schapira and Seuront [9] investigated how microscale nutrient patchiness is affected by turbulence. Based on a modeling approach, they found that phytoplankton exposed to high turbulence intensities are more efficient to uptake high concentration nitrogen pulses; on the other hand, uptake rates are higher for low concentration when turbulence is weak. Wagner et al. [10] investigated how scyphomedusae entrains and transports surrounding fluids and prey in order to catch calanoid copepods. The feeding currents generated by the medusa create a shear field that is well above the detection limit in copepods. However, only 58% of copepods reacted to the feeding currents. Hidden mechanisms in scale and flow fields may hinder the detection by copepods. Pécseli et al. [11] studied the encounter rate and the capture probabilities between cod larvae and prey (copepod) based on a field study under the different conditions of turbulence intensity.

Finally, it is essential to recognize and acknowledge the efforts provided by anonymous reviewers, which made it possible to maintain the high quality of all the contributions in this Special Issue.

Author Contributions: Both authors contributed equally to this article. All authors have read and agreed to the published version of the manuscript.

Conflicts of Interest: The authors declare no conflict of interest.

References

1. Tanaka, M. Changes in Vertical Distribution of Zooplankton under Wind-Induced Turbulence: A 36-Year Record. *Fluids* **2019**, *4*, 195. [CrossRef]
2. Niimoto, K.T.M.; Kuball, K.J.; Block, L.N.; Lenz, P.H.; Takagi, D. Rotational Maneuvers of Copepod Nauplii at Low Reynolds Number. *Fluids* **2020**, *5*, 78. [CrossRef]

3. Dabiri, J.O.; Colin, S.P.; Gemmell, B.J.; Lucas, K.N.; Leftwich, M.C.; Costello, J.H. Jellyfish and Fish Solve the Challenges of Turning Dynamics Similarly to Achieve High Maneuverability. *Fluids* **2020**, *5*, 106. [CrossRef]
4. Jiang, H. An Elastic Collision Model for Impulsive Jumping by Small Planktonic Organisms. *Fluids* **2020**, *5*, 154. [CrossRef]
5. Svetlichny, L.; Larsen, P.S.; Kiørboe, T. Kinematic and Dynamic Scaling of Copepod Swimming. *Fluids* **2020**, *5*, 68. [CrossRef]
6. Suwaki, C.H.; De-La-Cruz, L.T.; Lopes, R.M. Impacts of Microplastics on the Swimming Behavior of the Copepod *Temora turbinata* (Dana, 1849). *Fluids* **2020**, *5*, 103. [CrossRef]
7. Selander, E.; Fredriksson, S.T.; Arneborg, L. Chemical Signaling in the Turbulent Ocean—Hide and Seek at the Kolmogorov Scale. *Fluids* **2020**, *5*, 54. [CrossRef]
8. Borazjani, I. Numerical Simulations of Flow around Copepods: Challenges and Future Directions. *Fluids* **2020**, *5*, 52. [CrossRef]
9. Schapira, M.; Seuront, L. Nutrient Patchiness, Phytoplankton Surge-Uptake, and Turbulent History: A Theoretical Approach and Its Experimental Validation. *Fluids* **2020**, *5*, 80. [CrossRef]
10. Wagner, Z.; Costello, J.H.; Colin, S.P. Fluid and Predator-Prey Interactions of Scyphomedusae Fed Calanoid Copepods. *Fluids* **2020**, *5*, 60. [CrossRef]
11. Pécseli, H.L.; Trulsen, J.K.; Stiansen, J.E.; Sundby, S. Feeding of Plankton in a Turbulent Environment: A Comparison of Analytical and Observational Results Covering Also Strong Turbulence. *Fluids* **2020**, *5*, 37. [CrossRef]

© 2021 by the authors. Licensee MDPI, Basel, Switzerland. This article is an open access article distributed under the terms and conditions of the Creative Commons Attribution (CC BY) license (http://creativecommons.org/licenses/by/4.0/).

Communication

Changes in Vertical Distribution of Zooplankton under Wind-Induced Turbulence: A 36-Year Record

Mamoru Tanaka

Atmosphere and Ocean Research Institute, The University of Tokyo, Kashiwa, Chiba 277-8564, Japan; tanaka.mamoru0@gmail.com

Received: 17 October 2019; Accepted: 22 November 2019; Published: 25 November 2019

Abstract: A multidecadal record of a local zooplankton community, stored in an open-access database, was analyzed with wind data to examine the impact of wind-induced turbulence on vertical distribution of zooplankton. Two major findings were made. First, the abundance of zooplankton assemblage (composed of copepods, cladocerans, etc.) in the upper layer (<10 m deep) decreased with increasing turbulence intensity, suggesting turbulence avoidance by zooplankton. Second, when focusing on each species, it was found that ambush (sit-and-wait) feeders showed statistically significant changes in response to turbulence, whereas suspension (filter) feeders did not. This is the first clear evidence that ambush feeders change vertical distribution in response to turbulence.

Keywords: white sea; arctic ocean; net tow; turbulence avoidance; feeding mode; National Centers for Environmental Information; European Centre for Medium-Range Weather Forecasts

1. Introduction

While sub-centimeter-sized zooplankton play important roles in the marine ecosystem [1], processes that control their spatial distribution are elusive. Microscale turbulence, a ubiquitous characteristic of the ocean environment [2], significantly affects zooplankton swimming, feeding, and escape behavior [3]. Encounter rates with prey and mates are enhanced by environmental turbulence [4], but, at the same time, turbulence obscures signs of approaching predators, increasing the risks posed by staying in highly turbulent regions [5]. A numerical physical–ecological simulation, which considered the trade-off between reproduction and predation, suggested that avoidance of high levels of turbulence is most advantageous for reproduction [6]. Indeed, turbulence avoidance by zooplankton has been observed in small tank experiments, in which turbulence intensity was controlled by an oscillating grid [7]. However, field studies that demonstrate turbulence avoidance by zooplankton are limited [8–11]. Moreover, their conclusions are based on relatively short-term campaigns (<10 days) [8–11]. In this study, the impact of turbulence on zooplankton distribution was examined using a multidecadal record of a local zooplankton community and long-term sea-surface wind data.

2. Materials and Methods

2.1. Biological Parameters

Zooplankton data were obtained from an open-access database provided by the National Centers for Environmental Information. A biological dataset *"36-Year Time Series (1963–1998) of Zooplankton, Temperature, and Salinity in the White Sea"* [12] was used in this study. Zooplankton assemblage was sampled at the White Sea Biological Station (66°19.5' N, 33°39.4' E; the Arctic Ocean) from 1963 to 1998, i.e., 36 years. The water depth was 65 m at the station. A standard juday net was vertically towed every 10 days, collecting samples from surface (0 to 10 m deep), middle (10 to 25 m), and deep (25 to

65 m) layers. Mesh size and mouth area were 168 μm and 0.1 m^2, respectively. Zooplankton were fixed with 10% formaldehyde solution and classified to the species level. The net tows were performed during daytime. A total of 814 net tows were made.

Representative vertical position, mean depth distribution (MDD; m) was calculated for each net tow:

$$\text{MDD} = \frac{1}{N} \sum n_i z_i, \quad (1)$$

where N is the total abundance of zooplankton in the water column, n_i is the abundance in the ith depth bin (i.e., $i = 1, 2$, and 3), and z_i is the average depth of the ith depth bin (i.e., $z_1 = 5$ m, $z_2 = 17.5$ m, and $z_3 = 45$ m). MDD was calculated for the entire zooplankton assemblage, as well as for each species. In case a certain species was not observed in any depth layer, MDD was not calculated for this species. MDD and zooplankton abundance in the surface layer (individuals m^{-2}) were analyzed with turbulence intensity.

2.2. Physical Parameters

Turbulence intensity at the biological station was estimated with wind data provided by the European Centre for Medium-Range Weather Forecasts. The historical reanalysis dataset "ERA-40" [13] was used in this study, covering the entire period of the biological data. A sequence of wind speed at 10 m above the sea surface at the biological station was downloaded with a temporal resolution of 6 h. Wind events may need to be sustained for several hours to affect underwater distribution of zooplankton [8–11]. Hence, representative wind speeds U (m s^{-1}) during the net tows were obtained by averaging wind speeds at 09:00 and 15:00 in local time (U_{09} and U_{15}, respectively). The wind speed, U, was rejected when the difference between U_{09} and U_{15} exceeded half of U. Consequently, 105 of 814 data points (about 13%) were rejected.

Turbulent kinetic energy (TKE) dissipation rate ε (W kg^{-1}), which is a typical parameter to quantify turbulence intensity, was estimated by the "law of wall" method. This study employs an empirical equation valid for the layers shallower than 10 m deep, which was provided by MacKenzie and Leggett (1993) [14]:

$$\log_{10}(\varepsilon_{ML}) = a \cdot \log_{10}(U) + b \cdot \log_{10}(z) + c, \quad (2)$$

where ε_{ML} is the volume-based TKE dissipation rate (W m^{-3}) and z is the depth (m). The empirical coefficients were $a = 2.688$, $b = -1.322$, and $c = -4.812$, respectively [14]. As suggested by Equation (2), turbulence intensity due to wind greatly varies along the vertical coordinate. To determine the representative turbulence intensity for the surface layer (0 to 10 m), the ε_{ML} values were averaged over the surface layer, i.e., $\frac{1}{10} \sum_{z=1}^{10} \varepsilon_{ML}(z)$. Then, the averaged ε_{ML} (W m^{-3}) was converted to ε (W kg^{-1}) based on the typical seawater density of 1028 kg m^{-3} [2]. Since Equation (2) is not applicable for ice-covered periods (November to May), ε values were rejected for those periods (296 of 814 data points; 36%). Finally, 419 pairs of biological and physical parameters were analyzed.

2.3. Statistical Analyses

Long-term environmental changes, such as slow climate changes, could induce long-term trends both in the physical and biological parameters, resulting in spurious correlations between them [15]. Hence, such long-term trends were examined by linear regression models. Seasonal changes, which have the same problem, were also examined by analysis of variance (ANOVA) tests and autocorrelation analyses. The ANOVA tests were performed for the data divided into monthly intervals (i.e., June, July, August, September, and October), whereas the autocorrelation analyses were applied to monthly averaged data. As autocorrelation is applicable for equally spaced data points, the winter data (i.e., November to May), which were not used for any other analyses, were included in the autocorrelation analyses.

Then, potential effects of ε on surface abundance and MDD of the entire assemblage were examined by linear regression models and ANOVA tests. Logarithmic values of ε were used for the linear regression models. The ANOVA tests were performed for the data divided into 8 intervals of different turbulence levels, which were equally spaced in logarithmic scale. The relationships between ε vs. the surface abundance and MDD were also examined based on the datum averaged over each cycle (i.e., June to October) to eliminate potential bias due to seasonal changes. The effects of ε on each zooplankton species were examined by two-sample t-tests, where the data were divided into 2 levels of turbulence: one for low ($\varepsilon < 10^{-7}$ W kg^{-1}) and the other for high ($\varepsilon > 10^{-7}$ W kg^{-1}).

3. Results

Wind speed reached 10 m s^{-1} during the analysis period. TKE dissipation rate, ε, ranged from 10^{-8} to 10^{-6} W kg^{-1}. The average was $\varepsilon = 2 \times 10^{-7} \pm 2 \times 10^{-7}$ W kg^{-1} (mean ± standard deviation), typical in the surface layer [2]. Zooplankton assemblage was dominated by copepod species (Table 1), typical in the White Sea [16]. Abundance of entire zooplankton assemblage was highly concentrated in the surface layer (37,000 ± 46,000 individuals m^{-2}) relative to the middle and deep layers (17,000 ± 23,000 and 5000 ± 10,000 individuals m^{-2}, respectively). Average MDD of the entire assemblage was 12.9 ± 5.2 m.

Table 1. List of zooplankton species. Feeding mode definitions are based on the literature [17–19]. The effects of turbulence intensity on surface abundance (0 to 10 m deep) and mean depth distribution (MDD) were examined by two-sample *t*-tests, where the data were divided into two levels of turbulence: one for low ($\varepsilon < 10^{-7}$ W kg^{-1}) and the other for high ($\varepsilon > 10^{-7}$ W kg^{-1}). Boldface italics indicate statistical significance ($p < 0.02$).

			Surface Abundance (×10³ ind m⁻²)				MDD (m)			
					t-Test				t-Test	
	Feeding Mode	Turbulence Level	Mean ± Standard Error	df	t-Value	p-Value	Mean ± Standard Error	df	t-Value	p-Value
Copepod										
Acartia longiremis	Ambush/suspension	Low	3.13 ± 0.30	417	0.76	0.449	12.45 ± 0.58	405	−0.50	0.619
		High	2.87 ± 0.32				13.05 ± 0.60			
Microstella norvegica	Particle	Low	0.43 ± 0.10	417	1.06	0.290	13.47 ± 0.73	199	−0.17	0.864
		High	0.52 ± 0.11				13.17 ± 0.61			
Oithona similis	Ambush	Low	35.53 ± 2.19	417	3.50	**0.001**	11.98 ± 0.34	417	−2.55	**0.011**
		High	26.96 ± 1.96				13.10 ± 0.35			
Temora longicornis	Suspension	Low	7.36 ± 1.05	417	−0.29	0.771	13.36 ± 0.59	378	−1.99	0.047
		High	6.86 ± 1.08				14.54 ± 0.55			
Chaetognath										
Saggita elegans	Ambush	Low	0.53 ± 0.08	417	2.85	**0.005**	24.53 ± 0.85	401	−2.54	**0.011**
		High	0.42 ± 0.09				27.78 ± 0.84			
Cladoceran										
Evadne nordmanni	Suspension	Low	4.35 ± 0.57	417	1.71	0.088	7.64 ± 0.40	353	−0.97	0.335
		High	2.98 ± 0.42				7.83 ± 0.33			
Appendiclarian										
Fritillaria borealis	Suspension	Low	3.98 ± 0.55	417	1.42	0.156	9.68 ± 0.45	361	−1.35	0.179
		High	3.66 ± 0.83				10.66 ± 0.52			

Linear regression models showed that long-term trends were not statistically significant in ε ($p = 0.639$), surface abundance of entire assemblage ($p = 0.324$), nor MDD of the entire assemblage ($p = 0.822$) (Table 2). Yearly change in ε was -4×10^{-10} W kg^{-1} yr^{-1}, which corresponds to an overall decrease of 10^{-8} W kg^{-1} for 36 years (Figure 1a). This is one order smaller than the standard deviation of ε. Similarly, overall changes were calculated as +7380 individuals m^{-2} for the surface abundance and −0.2 m for the MDD (Figure 1b,c), much smaller than the standard deviations of those parameters. In contrast, the ANOVA tests showed that seasonal trends were significant for the surface abundance ($p < 0.001$) and the MDD ($p < 0.001$) (Table 3). The surface abundance has a peak in August, while the MDD increased from early summer to fall (Figure 2b,c). Such seasonal cycles are also suggested by the autocorrelation analyses (Figure 3b,c). Seasonal cycles in ε were found in the autocorrelation analysis (Figure 3a) but not in the ANOVA test ($p = 0.073$; Table 3; Figure 2a).

Table 2. Results of linear regression models to examine long-term trends in turbulent kinetic energy dissipation rate (denoted as "ε"), surface abundance of entire assemblage (0 to 10 m deep) ("Abundance"), and mean depth distribution of the entire assemblage ("MDD"). Data from Figure 1.

	n	$y = ax + b$ a	b	p-Value
ε	419	-4×10^{-10}	1×10^{-6}	0.639
Abundance	419	205	-4×10^{5}	0.324
MDD	419	-0.006	24	0.822

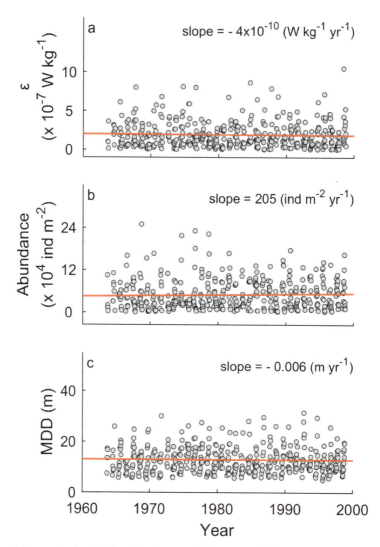

Figure 1. Time series for turbulent kinetic energy dissipation rate (ε) (**a**), surface abundance of the entire assemblage (0 to 10 m deep) (**b**), and mean depth distribution (MDD) of the entire assemblage (**c**). Filled circles denote raw data. Red lines denote linear regression models. Results of the regression models are summarized in Table 2.

Table 3. ANOVA table to examine the effects of month (denoted as "Month") on turbulent kinetic energy dissipation rate ("ε"), surface abundance of entire assemblage (0 to 10 m deep) ("Abundance"), and mean depth distribution ("MDD") of the entire assemblage. Data from Figure 2.

	Source	df	SS	MS	F	*p*-Value
ε	Month	4	3×10^{-13}	8×10^{-14}	2.2	0.073
Abundance	Month	4	2×10^{11}	6×10^{10}	42.7	<0.001
MDD	Month	4	3271	818	42.5	<0.001

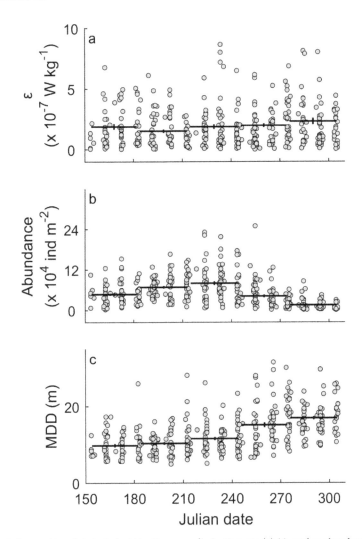

Figure 2. Seasonal trends in turbulent kinetic energy dissipation rate (ε) (**a**), surface abundance of the entire assemblage (0 to 10 m deep) (**b**), and mean depth distribution (MDD) of the entire assemblage (**c**). Filled circles denote raw data. Horizontal bars denote averages over month categories (i.e., June, July, August, September, and October). Error bars denote standard error. Results of the ANOVA tests are summarized in Table 3.

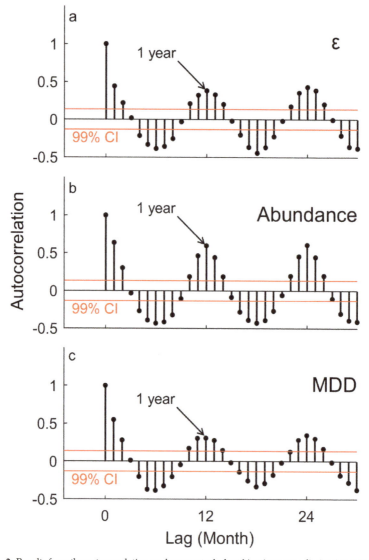

Figure 3. Results from the autocorrelation analyses on turbulent kinetic energy dissipation rate (ε) (**a**), surface abundance of the entire assemblage (0 to 10 m deep) (**b**), and mean depth distribution (MDD) of the entire assemblage (**c**). Red lines denote 99% confidence intervals.

The surface abundance of the entire assemblage decreased with increasing ε (Figure 4a). The linear regression model shows a correlation coefficient of r = −0.12 (p = 0.012) between the surface abundance and ε (Table 4). Additionally, the MDD increased (deepened) with increasing ε (Figure 4c). The regression model shows r = 0.14 (p = 0.003; Table 4) for the MDD. Such trends against ε can be clearly seen in bar graphs (Figure 4b,d). The ANOVA tests showed statistically significant differences among the different levels of ε (p = 0.004 for the surface abundance and p = 0.010 for the MDD; Table 4). The analyses for the data averaged over seasonal cycles also suggested negative and positive slopes for the surface abundance (r = −0.33) and the MDD (r = 0.29), respectively (Figure 5), consistent with those for the raw data (Figure 4a,c). However, those trends were not statistically significant (p = 0.052

for the surface abundance and $p = 0.093$ for the MDD; Figure 5). This is probably due to the small range of the average ε (Figure 5).

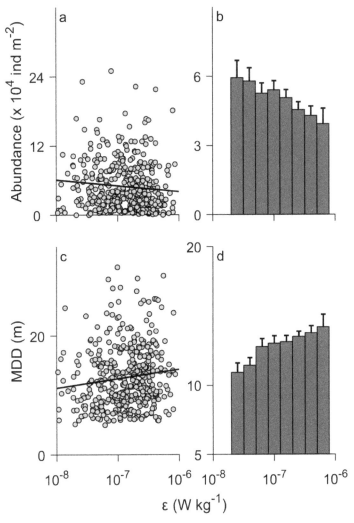

Figure 4. Surface abundance (0 to 10 m deep) (**a,b**) and mean depth distribution (MDD) (**c,d**) of the entire assemblage vs. turbulent kinetic energy dissipation rate (ε). The panels on the left show raw data. Each circle corresponds to a vertical net tow sample (n = 419). Solid lines denote linear regression models. The panels on the right show averages for different levels of turbulence. Error bars denote standard error. Results of the linear regression models and ANOVA tests are summarized in Table 4.

Table 4. Results of linear regression models and ANOVA tests to examine the effects of turbulent kinetic energy dissipation rate (denoted as "ε") on surface abundance of the entire assemblage (0 to 10 m deep) ("Abundance") and mean depth distribution ("MDD") of entire assemblage. r denotes the correlation coefficient. Data from Figure 4.

		Linear Regression Model					ANOVA				
		$y = ax + b$									
	n	a	b	r	p-Value	Source	df	SS	MS	F	p-Value
Abundance	419	-1×10^4	3×10^4	−0.12	0.012	ε	7	3×10^{10}	4×10^9	2.2	0.004
MDD	419	1.57	23.8	0.14	0.003	ε	7	510	73	2.7	0.010

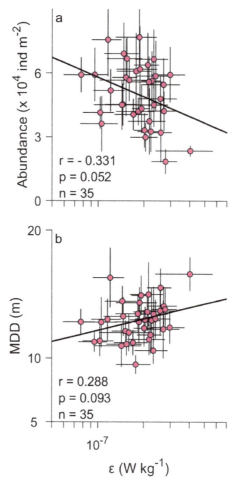

Figure 5. Same figures as in Figure 4 (left panels) but for the data averaged over seasonal cycle (i.e., June to October). Horizontal and vertical lines denote standard error. The first year, 1963, was excluded since net tows started from September. The number of samples is n = 35.

When focusing on each species, the feeding mode was found to be associated with sensitivity to turbulence. Ambush feeders, such as calanoid copepod *Oithona similis* and chaetognath *Saggita elegans*, showed a statistically significant decrease in surface abundance and increase in the MDD in response to increased ε ($p < 0.02$; Table 1). In contrast, suspension feeders, such as calanoid copepod

Temora longicornis, cladoceran *Evadne nordmanni*, and appendicularian *Fritillaria borealis*, showed no significant changes in surface abundance or MDD (Table 1). Calanoid copepod *Acartia longiremis*, which can switch between ambush and suspension feeding, exhibited no significant changes (Table 1). No changes were found for harpacticoid copepod *Microstella norvegica*, a particle feeder.

4. Discussion

Analysis of a multidecadal record of a local zooplankton community revealed avoidance of wind-induced turbulence by zooplankton, whereas significant response to turbulence was found only in ambush feeders (Table 1). This is consistent with laboratory experiments that demonstrated that ambush feeding is hindered by high levels of turbulence, while suspension feeding is less dependent on turbulence intensity [20,21]. Such turbulent effects on ambush feeding are also demonstrated by a theoretical model, which is designed to predict gut contents of ambush feeders in turbulent water [22]. In the literature, the model results were compared with those from field campaigns and showed that gut contents of ambush feeders decreased with increasing turbulence intensity [22]. Results from this study are consistent with those from the experimental and theoretical works [20–22].

In contrast to ambush feeders, pure suspension feeders exhibited no significant changes in response to turbulence. Given that suspension feeders are able to adapt to relatively high levels of turbulence [6], they probably place less priority on changing their position. Additionally, *Acartia longiremis*, which have multiple feeding modes, may switch their feeding mode to suspension feeding when in turbulent waters, rather than seek out low levels of turbulence (a similar discussion is seen in [21]). Although the particle feeder *Microstella norvegica* showed no significant changes in response to turbulence, this species, in another study, exhibited significant migration to deeper depths in response to wind-induced turbulence [11]. The reason for the difference between this study and the literature [11] remains unclear.

Physical processes, such as wind [23] and surface cooling [24], frequently disturb surface waters, producing vertical gradients in turbulence intensity in the water column [25]. Hence, downward migration is generally optimal behavior to seek lower levels of turbulence. However, turbulence is also generated by other processes, such as bottom stress associated with barotropic tides and swells [26], and shear stress associated with internal gravity waves [27]. Hence, actual turbulence intensities in the ocean would be different from those estimated by the simple model in Equation (2). This will result in errors in estimation of TKE dissipation rate (ε). Additionally, the reanalysis dataset would include potential errors in wind speed, which consequently induce additional errors in ε The European Centre for Medium-Range Weather Forecasts does not quantify the magnitude of the errors, but it could be substantial.

The r value of surface abundance vs. ε was r = −0.12 (Table 4), which corresponds to the coefficient of determination, r^2, of 0.0144. This means only 1% of the fluctuation is explained by ε. Zooplankton generally exhibit highly intermittent distributions in the ocean, resulting in high levels of inter-sample variability in zooplankton density [28]. This means that single/instantaneous samples could have substantial (generally large) differences from the true density [28]. Such an inter-sample variability would be a source of potential errors (or biological noise) in surface abundance (Figures 1b, 2b and 4a).

Here, I compare turbulent velocity to typical swimming speeds of zooplankton. While turbulent flow speeds are highly variable in space and time, the representative velocity scale near the boundary is the friction velocity (u_*) (m s^{-1}). The friction velocity (u_*) is a function of ε and the depth (z) (m) (e.g., Equation (4.9) in [2]):

$$u_* = (\varepsilon\, z\, \kappa)^{1/3}, \qquad (3)$$

where κ is the von Karman constant (=0.41). Assuming $\varepsilon = 10^{-7}$ W kg^{-1} (at which the significant changes in surface abundance and MDD were found) and $z = 10$ m, we obtain $u_* = 0.7$ cm s^{-1}, which is on the same order as the turbulent velocity scales in the seasonal thermocline [29]. This is comparable with typical swimming speeds of sub-centimeter-sized zooplankton, but much lower than their instantaneous escape jumps >10 cm s^{-1} [29,30]. Hence, I suspect that they can perform oriented swimming even in highly turbulent waters to seek optimal levels of turbulence.

Despite the potential errors inherent in the open-access database, statistically significant trends were found between turbulence and ambush feeders. Turbulence estimation based on wind speed is valid for the surface layer and provides information about physical processes at scales of individual plankton, which is generally absent in biological samplings. I hope this study encourages other researchers to examine the reproducibility of the observed trends.

Funding: This research received no external funding.

Acknowledgments: The author gratefully acknowledges discussions with Amatzia Genin, Rubens M. Lopes, Gregory N. Ivey, Yoshinari Endo, J. Rudi Strickler, and Hidekatsu Yamazaki.

Conflicts of Interest: The author declares no conflicts of interest.

References

1. Steinberg, D.K.; Landry, M.R. Zooplankton and the ocean carbon cycle. *Annu. Rev. Mar. Sci.* **2017**, *9*, 413–444. [CrossRef] [PubMed]
2. Thorpe, S.A. *The Turbulent Ocean*; Cambridge University Press: Cambridge, UK, 2005.
3. Kiørboe, T. *A Mechanistic Approach to Plankton Ecology*; Princeton University Press: Princeton, NJ, USA, 2008.
4. Yamazaki, H.; Osborn, T.R.; Squires, K.D. Direct numerical simulation of planktonic contact in turbulent flow. *J. Plankton Res.* **1991**, *13*, 629–643. [CrossRef]
5. Gilbert, O.M.; Buskey, E.J. Turbulence decreases the hydrodynamic predator sensing ability of the calanoid copepod *Acartia tonsa*. *J. Plankton Res.* **2005**, *27*, 1067–1071. [CrossRef]
6. Visser, A.W.; Mariani, P.; Pigolotti, S. Swimming in turbulence: Zooplankton fitness in terms of foraging efficiency and predation risk. *J. Plankton Res.* **2009**, *31*, 121–133. [CrossRef]
7. Seuront, L.; Yamazaki, H.; Souissi, S. Hydrodynamic disturbance and zooplankton swimming behavior. *Zool. Stud.* **2004**, *43*, 376–387.
8. Lagadeuc, Y.; Boulé, M.; Dodson, J.J. Effect of vertical mixing on the vertical distribution of copepods in coastal waters. *J. Plankton Res.* **1997**, *19*, 1183–1204. [CrossRef]
9. Incze, L.S.; Hebert, D.; Wolff, N.; Oakey, N.; Dye, D. Changes in copepod distributions associated with increased turbulence from wind stress. *Mar. Ecol. Prog. Ser.* **2001**, *213*, 229–240. [CrossRef]
10. Visser, A.W.; Saito, H.; Saiz, E.; Kiørboe, T. Observations of copepod feeding and vertical distribution under natural turbulent conditions in the North Sea. *Mar. Biol.* **2001**, *138*, 1011–1019. [CrossRef]
11. Maar, M.; Visser, A.W.; Nielsen, T.G.; Stips, A.; Saito, H. Turbulence and feeding behaviour affect the vertical distributions of *Oithona similis* and *Microsetella norwegica*. *Mar. Ecol. Prog. Ser.* **2006**, *313*, 157–172. [CrossRef]
12. 36-Year Time Series (1963–1998) of Zooplankton, Temperature, and Salinity in the White Sea. Available online: https://www.nodc.noaa.gov/OC5/WH_SEA/index1.html (accessed on 20 November 2019).
13. The ERA-40 archive. Available online: https://www.ecmwf.int/node/10595 (accessed on 20 November 2019).
14. MacKenzie, B.R.; Leggett, W.C. Wind-based models for estimating the dissipation rates of turbulent energy in aquatic environments: Empirical comparisons. *Mar. Ecol. Prog. Ser.* **1993**, *94*, 207–216. [CrossRef]
15. Prairie, Y.T.; Bird, D.F. Some misconceptions about the spurious correlation problem in the ecological literature. *Oecologia* **1989**, *81*, 285–288. [CrossRef] [PubMed]
16. Pertsova, N.M.; Kosobokova, K.N. Zooplankton of the White Sea: Features of the composition and structure, seasonal dynamics, and the contribution to the formation of matter fluxes. *Oceanol. C/C Okeanol.* **2003**, *43*, S108–S122.
17. Flood, P.R. House formation and feeding behaviour of *Fritillaria borealis* (Appendicularia: Tunicata). *Mar. Biol.* **2003**, *143*, 467–475. [CrossRef]
18. Tönnesson, K.; Tiselius, P. Diet of the chaetognaths *Sagitta setosa* and *S. elegans* in relation to prey abundance and vertical distribution. *Mar. Ecol. Prog. Ser.* **2005**, *289*, 177–190. [CrossRef]
19. Brun, P.; Payne, M.R.; Kiørboe, T. A trait database for marine copepods. *Earth Syst. Sci. Data* **2017**, *9*, 99–113. [CrossRef]
20. Saiz, E.; Kiørboe, T. Predatory and suspension feeding of the copepod *Acartia tonsa* in turbulent environments. *Mar. Ecol. Prog. Ser.* **1995**, *122*, 147–158. [CrossRef]

21. Saiz, E.; Calbet, A.; Broglio, E. Effects of small-scale turbulence on copepods: The case of *Oithona davisae*. *Limnol. Oceanogr.* **2003**, *48*, 1304–1311. [CrossRef]
22. Pécseli, H.L.; Trulsen, J.K.; Stiansen, J.E.; Sundby, S.; Fossum, P. Feeing of plankton in turbulent oceans and lakes. *Limnol. Oceanogr.* **2019**, *64*, 1034–1046. [CrossRef]
23. Oakey, N.S.; Elliott, J.A. Dissipation within the surface mixed layer. *J. Phys. Oceanogr.* **1982**, *12*, 171–185. [CrossRef]
24. Shay, T.J.; Gregg, M.C. Convectively driven turbulent mixing in the upper ocean. *J. Phys. Oceanogr.* **1986**, *16*, 1777–1798. [CrossRef]
25. Smyth, W.D.; Moum, J.N. 3D Turbulence. In *Encyclopedia of Ocean Sciences*, 3rd ed.; Cochran, J.K., Bokuniewicz, H.J., Yager, P.L., Eds.; Elsevier Ltd.: London, UK, 2019; Volume 3, pp. 486–496.
26. Drost, E.J.F.; Lowe, R.J.; Ivey, G.N.; Jones, N.L. Wave-current interactions in the continental shelf bottom boundary layer of the Australian North West Shelf during tropical cyclone conditions. *Cont. Shelf Res.* **2018**, *165*, 78–92. [CrossRef]
27. Kokubu, Y.; Yamazaki, H.; Nagai, T.; Gross, E.S. Mixing observations at a constricted channel of a semi-closed estuary: Tokyo Bay. *Cont. Shelf Res.* **2013**, *69*, 1–16. [CrossRef]
28. Downing, J.A.; Pérusse, M.; Frenette, Y. Effect of interreplicate variance on zooplankton sampling design and data analysis. *Limnol. Oceanogr.* **1987**, *32*, 673–680. [CrossRef]
29. Yamazaki, H.; Squires, K.D. Comparison of oceanic turbulence and copepod swimming. *Mar. Ecol. Prog. Ser.* **1996**, *144*, 299–301. [CrossRef]
30. Buskey, E.J.; Swift, E. Behavioral responses of oceanic zooplankton to simulated bioluminescence. *Biol. Bull.* **1985**, *168*, 263–275. [CrossRef]

 © 2019 by the author. Licensee MDPI, Basel, Switzerland. This article is an open access article distributed under the terms and conditions of the Creative Commons Attribution (CC BY) license (http://creativecommons.org/licenses/by/4.0/).

Article

Rotational Maneuvers of Copepod Nauplii at Low Reynolds Number

Kacie T. M. Niimoto [1,2], Kyleigh J. Kuball [1], Lauren N. Block [1], Petra H. Lenz [1] and Daisuke Takagi [1,2,3,*]

[1] Bekesy Laboratory of Neurobiology, Pacific Biosciences Research Center, University of Hawaii at Manoa, Honolulu, HI 96822, USA; ktmn@hawaii.edu (K.T.M.N.); kyleighk@hawaii.edu (K.J.K.); blockln7@hawaii.edu (L.N.B.); petra@hawaii.edu (P.H.L.)
[2] Department of Mechanical Engineering, University of Hawaii at Manoa, Honolulu, HI 96822, USA
[3] Department of Mathematics, University of Hawaii at Manoa, Honolulu, HI 96822, USA
* Correspondence: dtakagi@hawaii.edu

Received: 1 April 2020; Accepted: 18 May 2020; Published: 21 May 2020

Abstract: Copepods are agile microcrustaceans that are capable of maneuvering freely in water. However, the physical mechanisms driving their rotational motion are not entirely clear in small larvae (nauplii). Here we report high-speed video observations of copepod nauplii performing acrobatic feats with three pairs of appendages. Our results show rotations about three principal axes of the body: yaw, roll, and pitch. The yaw rotation turns the body to one side and results in a circular swimming path. The roll rotation consists of the body spiraling around a nearly linear path, similar to an aileron roll of an airplane. We interpret the yaw and roll rotations to be facilitated by appendage pronation or supination. The pitch rotation consists of flipping on the spot in a maneuver that resembles a backflip somersault. The pitch rotation involved tail bending and was not observed in the earliest stages of nauplii. The maneuvering strategies adopted by plankton may inspire the design of microscopic robots, equipped with suitable controls for reorienting autonomously in three dimensions.

Keywords: locomotion; reorientation; swimming microorganism

1. Introduction

Animals change the orientation of their body by coordinating the movements of various body parts. For animals capable of moving quickly with considerable inertia through air or water, they may rotate easily with minor body adjustments, as demonstrated by fruit flies [1] and spinner dolphins [2]. For microscopic organisms with small inertia, however, they must actively and repeatedly move their body parts in order to rotate adequately in fluids dominated by viscosity. In this physical regime of low Reynolds number (Re), it is well known that bacteria can tumble [3], and phototactic algae can reorient [4], using flexible flagella. However, microcrustaceans such as larval copepods have relatively stiff bodies and appendages. They have been observed to turn sharply, but their rotational motion is not as well understood as their translational motion [5,6]. Thus, further research is needed to unravel the physical mechanisms underlying the rotational motion of larval copepods.

Numerous theoretical and physical models have considered the problem of reorientation with rigid body parts at low Reynolds number. A key constraint on bodies with minimal inertia is that they cannot translate or rotate by themselves through movements that are reversible in time [7]. Cycles of irreversible kinematic changes are needed to reorient, and this requires at least two degrees of freedom. One of the simplest models of a reorienting body consists of three spheres, each connected to a rigid rod of equal length, with the rods meeting at a common point [8]. The whole system can rotate through an irreversible sequence of cyclical changes in the angles between the rods. Other examples of reorienting bodies include three rigid spheres that are arranged in a triangle and connected by springs, which are stretched and compressed in sequential order [9], and a pair of paddles, fitted with disks at each end, driven to rotate, with adjustable spacing between the paddles [10]. These simple swimmers have elucidated the minimal components needed for reorienting. Still, their maneuverability is limited and inefficient because many cycles of body movements are typically needed, e.g., to reverse the body's orientation. Practical applications of microscopic robots may require rapid reorientations. Previous experiments have demonstrated that the robots can be steered by exerting a torque with electric and magnetic fields [11,12]. However, these robots are steered and controlled externally; a fully autonomous microrobot must be capable of responding adequately to changes in its local environment, with minimal external force or torque.

Here we turn to nature for inspiration, focusing on the rotational maneuvers of copepods, a common group of zooplankton thriving in the world's oceans and lakes. Maneuverability is crucial to the survival of adult copepods and their offspring (nauplii). A diversity of behaviors has been described in nauplii, which range from periods of immobility, swimming associated with foraging, and escape swims. Tracks of swimming trajectories in three dimensions have shown that, over time intervals in the range of minutes, nauplii typically move in helical patterns, using a stop-and-go pattern [13–16]. These authors also reported species-specific patterns, suggesting significant flexibility in behavior, despite the similarity in design. Copepods are under extreme predatory pressure from many aquatic organisms [17–19], and consequently, they have evolved remarkable escape responses to predatory attacks [15,20–24]. These responses often require a rapid reorientation to avoid swimming into the mouth of the predator [25]. Nauplii operate at lower Reynolds numbers than adults because of their smaller size, and they have a challenging task of capturing prey. While in later developmental stages, copepodids generate feeding currents, relatively few nauplii are able to do so [14]. Instead, nauplii swim toward their prey, generating a bow wave. In order to succeed in capturing the prey, a nauplius encircles the prey to draw it to the mouth while maneuvering its appendages [26]. While it is known that three pairs of appendages are involved, the details of appendage movements have not been resolved. Crustacean nauplii have three pairs of appendages that beat at a range of frequencies, with maximum frequencies recorded in copepod nauplii (>100 Hz; [27]). The asynchronous beating of appendages has been recognized as a factor for translating the body [5,6,28,29], and asymmetry in the appendage movement has been identified as a factor for reorienting the body [26]. However, the causal relationship between the movements of the body and the appendages and the resulting orientation remains unclear. Thus, the basic physical mechanisms behind the maneuvers have not yet been identified.

This work characterized the rotation of copepod nauplii around three perpendicular axes, as defined with respect to their body. For the rotation around each axis, we observed a sequence of movements of the appendages. Additionally, the tail moved considerably during the rotation around one of the axes. We interpreted these observations by using basic physical arguments and identified plausible mechanisms driving the rotation around each axis. Our findings suggest that, as the nauplius grows and develops a more pronounced tail, it gains an additional and efficient way of reorienting by performing a backflip somersault. Without the ability to rotate directly around one axis, the earliest stages of nauplii would need a strategic combination of rotations around the other two axes, in order to maneuver in three-dimensional space.

2. Materials and Methods

2.1. High-Speed Videography of Nauplii Swimming

Two species of Paracalanid copepods, *Bestiolina similis* and *Parvocalanus crassirostris*, were isolated from Kaneohe Bay (Oahu, Hawaii) and cultured by using the methods described in a previous study [30]. Temperature and salinity were maintained between 23 and 25 °C and approximately 35 ppt, within the typical conditions of Kaneohe Bay [31]. The two species are comparable in size, developmental progression, and behaviors. They move intermittently in three dimensions. The nauplii (70–150 µm in size) were isolated from the cultures and gently pipetted into videography containers containing seawater and low densities of residual *Tisochrysis lutea*. We used two different methods for observing the nauplii: one from the side and another from above. The view from above was magnified by using 10× and 40× objective lenses of an inverted microscope (Olympus IX73). A small Petri dish (54.5 mm diameter) containing the nauplii was shifted horizontally, until a nauplius appeared in the field of view of the microscope. The second apparatus consisted of a cuvette (10 × 10 × 45 mm) containing the nauplii viewed from the side and magnified by comparable amounts, using an objective and condenser lens. We observed similar behavior and results from the side as from above. All videos were recorded with a high-speed camera (Phantom Miro M110), at frame rates between 1000 and 2000 fps; a total of 95 maneuvers were recorded.

2.2. Video Analysis

The videos were analyzed by using ImageJ software. Rotational maneuvers were analyzed frame by frame, and the rotations were categorized into three general types: yaw, roll, and pitch. The orientation of the body was determined by observing the plane formed by the appendages, the location of the labrum, and the direction of the curved tail. The labrum, or mouthparts of the nauplii, are located ventrally, and the tail curves toward the labrum. The appendage plane gives a sense of how the ventral/dorsal sides are oriented, while the labrum and curvature of the tail help to identify the ventral side of the body. The orientation could also be determined by tracking the motion of the nauplius throughout its maneuver and observing any appendage crossover during the rotation. Selected videos were analyzed by using Tracker and Microsoft Excel software. Tracker was used to track the body's midpoint and the endpoint of each appendage. The change of appendage angles with respect to the nauplius's initial heading direction was computed in Excel.

3. Results

3.1. Overview

Nauplii of *B. similis* and *P. crassirostris* were observed to rotate around three different axes. Figure 1 shows the axes and general anatomy of the nauplius. Nauplii have three pairs of appendages: antennules (A1), antennae (A2), and mandibles (M). The axes were defined such that the roll axis is along the length of the body; the pitch axis extends across the body, toward the right-hand side of the body, as perceived from the body; and the yaw axis is perpendicular to the plane formed by the pitch and roll axes, pointing in the direction from the ventral to the dorsal side of the body. During swimming, the appendages oscillate approximately in the plane containing the roll and pitch axes, at a frequency on the order of f~100 Hz, corresponding to a duration of T~10 milliseconds every periodic cycle. The appendages generate fluid flow at Reynolds number Re = L(L/T)/ν~0.3, where ν~1 mm^2 s^{-1} is the kinematic viscosity of water at room temperature, L~0.05 mm is the length scale set by the average length of all appendages, and L/T is the velocity scale. The Reynolds number is less than 1, meaning that the rotational motion of the nauplii is governed primarily by viscosity and less by inertia.

Our results depend greatly on the developmental stage of copepod nauplii. There are six nauplius stages (N1–N6); the first stage (N1) is the smallest and has a nearly spherical shape. The body becomes larger and more elongated in shape as it develops into later stages. Of the six nauplius stages, the first two stages, N1 and N2, are typically non-feeding [32]. The nauplii begin feeding at the N3 stage, which is distinguished from N1/N2 by the development of a noticeable tail. The nauplius stages were identified by using the shape and the length of the body and categorized as non-feeding (N1/N2) and feeding (N3–N6). The feeding stages are referenced as ≥ N3 throughout the rest of this paper. The relative occurrences of all three rotations are shown in Table 1. Some videos showed rotations around axes which were unclear; these were not included for simplification. Rotations about the yaw and pitch axis by an angle less than 45° and roll rotations by an angle less than 90° were considered incomplete and were not counted in the dataset. The significance of the table is that the rotation about the pitch axis was not observed at the early stages (N1/N2), whereas the later stages (≥N3) displayed rotations about all three axes. Despite the apparent limitations of N1/N2, the nauplii are capable of exploring in three-dimensional space, as we discuss further below. We first describe how the body moves during each rotation, starting with the yaw rotation.

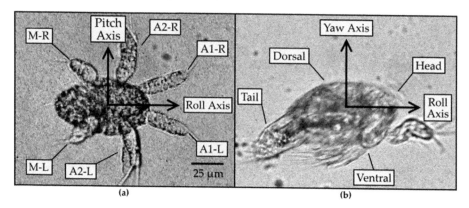

Figure 1. (a) Image of an N2 nauplius with the ventral side in view at 40× magnification. The three pairs of appendages (antennules (A1), antennae (A2), and mandibles (M)) are each labeled L or R on the left or right side of the body, respectively. The pitch axis is defined to point in the direction from the left to the right side of the body; (b) image of ≥N3 nauplius at 40× magnification. The roll axis points in the direction from the tail to the head end of the body. The yaw axis points from the ventral to the dorsal side of the body.

Table 1. Yaw, roll, and pitch occurrences by naupliar stage.

	Yaw	Roll	Pitch
N1/N2	10	21	0
≥N3	15	19	30

3.2. Yaw Rotation

Figure 2 shows a time series and trajectory of a typical yaw rotation, which involves rotating around the yaw axis and turning toward one side of the body (see Supplementary Video 1). During this rotation, the midpoint of the body undergoes considerable back-and-forth translation in the plane perpendicular to the yaw axis. The net displacement of the body was measured to be 240 ± 100 μm after rotating 180° around the yaw axis (mean and standard deviation, n = 5). The body alternates between moving forward (blue) and backward (red) along directions pointing approximately along the roll axis of the body. The directions of forward and subsequent backward motion are not precisely parallel; instead, they are offset systematically by a small angle, which leads to substantial turning after multiple cycles. The rotation efficiency was defined as the angular change in the orientation of the body every periodic cycle, and this was measured to be 33 ± 6° for yaw rotation (mean and standard deviation, n = 5).

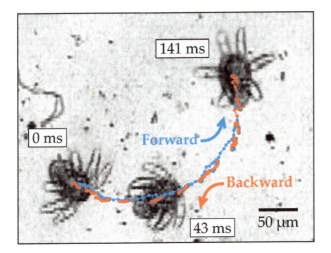

Figure 2. Time series of an N1/N2 nauplius undergoing yaw rotation observed at 10× magnification. The three snapshots of the nauplius were taken at different times, as indicated. The trajectory of the body's midpoint is shown by dots, which are colored blue or red, depending on the direction of motion of the body.

To identify any left–right asymmetry in the movement of the appendages, we tracked the orientation of the body and all six appendages during the yaw rotation (Figure 3). The orientation of the body relative to its original orientation shows noticeable oscillations with time (Figure 3a). However, there is no apparent difference between the orientation of the appendages on the left and right sides of the body (Figure 3b).

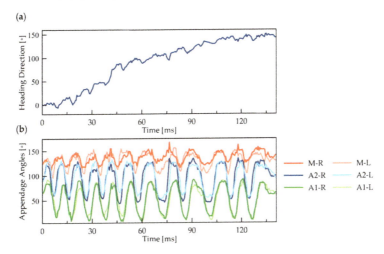

Figure 3. Yaw rotation of ≥N3 nauplius: (**a**) cumulative rotation over time of the body's heading direction (roll axis) around the yaw axis; (**b**) temporal change in the orientation of six appendages, measured by the angle with respect to the body's roll axis. Orange, blue, and green curves represent the M, A2, and A1 pairs, respectively. The light and dark colors correspond to the left and right sides.

To gain insight into possible mechanisms underlying the yaw rotation, we observed the nauplii at higher (40×) magnification, as shown in Figure 4 and Supplementary Video 2. Figure 4a reveals the setae, flexible hair-like structures protruding from the distal tips of each appendage. These setae might fold or expand to different degrees on either side of the body, which would offer a possible mechanism for turning. However, we found no clear evidence of such asymmetry in the setae movement. Instead, we observed a noticeable asymmetry in the oscillation of the A2 appendages on the left and right sides of the body during the yaw rotation (Figure 4b). This was noticeable because the A2 appendage is biramous and splits into two branches. The two branches can appear to overlap if one lies above the other branch, as viewed in the direction parallel to the yaw axis. Alternatively, they appear as distinct branches if the appendage rotates by an angle close to 90 degrees around its long axis. This type of rotation is referred to as pronation or supination, depending on whether the branch on the ventral side swings toward or away from the body, respectively. The difference between pronation and supination was difficult to distinguish in our videos. Furthermore, the angle of rotation around the long axis of the appendage was difficult to quantify. Nevertheless, the two branches became more visible on the left but not on the right side of the body, indicating that the left appendage either pronated or supinated, while the right appendage did not. This implies that the rotated A2 appendage has a smaller profile area and thus drives less of the surrounding fluid on the left than the right during the important power stroke, which offers a possible mechanism for swerving the body to the left, as observed.

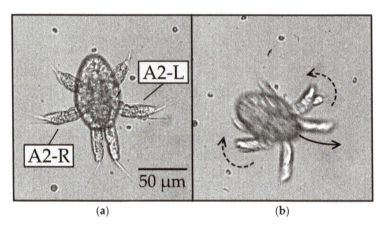

Figure 4. Yaw rotation of N1/N2 nauplius with the dorsal side in view at 40× magnification: (**a**) at the start of the rotation, the A2 appendages have branches appearing to overlap (0 ms); (**b**) as the body swerves to the left (counterclockwise), the two branches of the biramous A2 appendage become more visible in the left compared to the right A2 (21 ms).

3.3. Roll Rotation

Next, we describe the roll rotation, which involves the body rotating around the roll (main) axis while swimming along the axis. The body follows a helical path in a near-linear direction. The net displacement of the body after rotating around the roll axis by 180 degrees was measured to be 800 ± 120 µm (mean and standard deviation; n = 5), which is considerably greater than the net displacement of the body following yaw rotation, as described earlier. During the roll rotation, the body alternates repeatedly between forward and backward displacements, which are comparable to those measured previously during linear swimming without rolling [6]. The large displacement of the body during the roll rotation makes it more challenging to observe the nauplii, because they only remain in focus for a brief period when they swim in and out of the field of view. Figure 5 shows a typical image sequence of a nauplius rotating around the roll axis by approximately 180° (see Supplementary Video 3). The right side of the body is initially in view, followed by the ventral side and then the left side. The rotation efficiency of the roll rotation was measured to be 27 ± 6°, which is comparable to that of the yaw rotation described earlier.

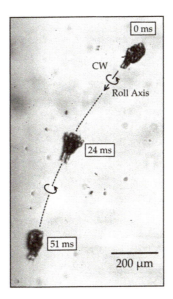

Figure 5. Time series of a roll rotation. Initially (0 ms), the right side of the body is in view, then the ventral side (24 ms), and then the left side of the body (51 ms).

The physical mechanism driving the roll rotation was difficult to visualize because of the challenges associated with observing the moving appendages in sufficient detail. A magnified view of the roll rotation showed that the appendages alternate between power and return strokes (see Supplementary Video 4), similar to appendage movements during the yaw rotation, as described earlier. There was no apparent explanation for how the roll rotation around the swimming direction was produced. Unlike the flexible cilia and flagella that are known to generate helical motion in swimming microorganisms [33,34], the appendages of copepod nauplii are rigid. One possible mechanism is that the tip of an appendage undergoes orbital motion as opposed to tracing the same curved path back and forth. This could repeatedly drive the surrounding fluid around the roll axis, though we observed no clear evidence of such orbital motion of appendages in our videos. Another possible mechanism is that an appendage pronates or supinates by an acute angle, which was more evident (see Supplementary Video 4). If an appendage remains pronated or supinated by an acute angle, say 45 degrees, during the power stroke, the appendage could drive the surrounding fluid around the roll axis and thereby resolve how the nauplii produce the roll rotation. The key is to produce left–right asymmetry, e.g., with pronation of the A2 appendage on the left or right side, but not both. This behavior was observed in all nauplian stages, including the early non-feeding stages.

3.4. Pitch Rotation

The pitch rotation produced relatively little translation of the body in three dimensions, enabling us to readily observe the rotation while the body remained in focus. The net displacement of the body after rotating around the pitch axis by 180 degrees was measured at 60 ± 50 μm (mean and standard deviation; n = 5), with the smallest being 15 μm in the plane of the field of view. Figure 6 shows a typical image sequence of the rotation around the pitch axis by approximately 180° (see Supplementary Video 5). In the first image, the ventral side is in view, and the roll axis points toward the bottom-left corner of the image. In the second image, the tail end is in view, confirmed by the appendages on the ventral side appearing in the bottom-left corner of the image. Contrarily, if the head end were in view, the appendages would be located toward the top-right corner. In the third and final image, the dorsal side is in view. Thus, the body has flipped over by performing a backflip somersault. The rotational efficiency of the pitch rotation was measured to be 60 ± 30 ° (mean and standard deviation; n = 5), which is much higher than those of the yaw and roll rotations presented earlier.

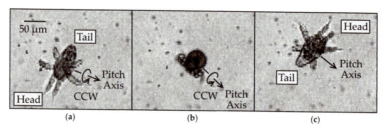

Figure 6. Image sequence of pitch rotation: (**a**) ≥N3 nauplius with ventral side in view (time t = 0); (**b**) tail end in view (t = 10 ms); (**c**) dorsal side in view (t = 22 ms).

Observations from one side of the nauplius along the pitch axis provided additional insights into the possible mechanism underlying the pitch rotation (Figure 7). The pitch rotation consists of the following sequence: contraction of the tail, contraction of the appendages, expansion of the tail, and expansion of the appendages, in this particular order.

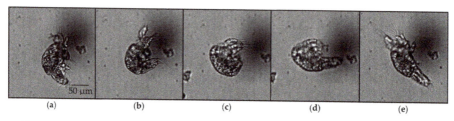

Figure 7. Image sequence of an ≥N3 nauplius undergoing pitch rotation, as viewed in the plane perpendicular to the pitch axis: (**a**) tail and appendages are expanded (0 ms); (**b**) appendages remained expanded, and tail was contracted (21 ms); (**c**) tail remained contracted, and appendages contracted (34 ms); (**d**) appendages remained contracted, and tail expanded (40 ms); (**e**) tail remained expanded, and appendages expanded (48 ms).

To understand why this particular appendage-and-tail-movement sequence results in the rotation of the body around the pitch axis, we consider a simplified theoretical model consisting of three slender rods connected in series by hinges, as sketched in Figure 8. Previous studies focused on the translation of the body, commonly known as Purcell's swimmer [7,35]. Here we consider the rotation of the body. For simplicity, the six appendages are grouped into one end of the body, and the tail at the other. The rods do not incorporate the shape and movement of the tail and appendages in any quantitative detail, but they are sufficient for elucidating the basic effects of contracting and expanding the body parts in the particular sequence described before.

Suppose the two rods at the ends, hereafter referred to as the tail and arm of the body, are nearly parallel to the central rod initially (Figure 8a). First, the angle between the central rod and the tail is assumed to contract and close (Figure 8b). By conservation of angular momentum, the central rod rotates in the opposite direction to the tail, though the rotation angle is smaller in magnitude because the central rod experiences more drag than the relatively short tail. Second, the angle between the arm and the central rod is assumed to contract and close (Figure 8c). The arm and the central rod rotate in opposite directions by conservation of angular momentum again. However, the rotation angle of the arm is less than that of the tail in the previous step, because of the contracted tail. Third, the tail swings back open (Figure 8d). The rotation angle of the tail is equal to that of the arm in the previous step by symmetry. Fourth, the arm swings back open (Figure 8e). This has correspondence with the time-reversal of closing the tail in the first step. After this entire sequence, the body rotates in the direction consistent with the experimental observations described earlier. Note that, without the ability to bend the tail, the simplified model would have only one hinge instead of two hinges. Such a body with only one degree of freedom would not be able to rotate at a low Reynolds number, which is consistent with the lack of observation of pitch rotations in N1/N2 nauplii.

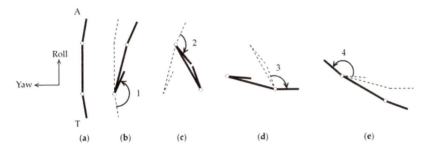

Figure 8. Simplified theoretical model for pitch rotation: (**a**) initial configuration of a body equipped with an arm (A) and a tail (T); (**b**) the tail closes and rotates by a large angle (1); the body rotates by a relatively small angle in the opposite direction. Dashed lines show the earlier configuration for comparison; (**c**) the arm closes and rotates by a moderate angle (2); (**d**) the tail opens and rotates by a moderate angle (3); (**e**) the tail opens and rotates by a large angle (4).

3.5. Direction of Rotation

The direction of rotation of the body was categorized into either clockwise (CW) or counterclockwise (CCW), according to the right-hand rule: CCW rotation around an axis is defined such that, if the right thumb points in the direction of the axis, as defined in Figure 1, then the right fingers curl in the same direction as the reorienting body. For example, a CCW yaw corresponds to a nauplius with the dorsal side in view, appearing to swerve to the left side of the body. A CCW roll corresponds to the body, with the head end in view, appearing to rotate around the roll (main) axis in the CCW direction. A CCW pitch corresponds to the body with the right side in view, appearing to perform a backflip.

Table 2 shows that the nauplii are capable of rotating in both directions around the yaw and roll axes. The number of observations in Table 2 is less than that in Table 1 because the ventral and dorsal sides were unclear in some videos, and thus the direction of rotation could not be determined in some yaw and roll rotations. Nonetheless, the result of pitch rotation is significant. The nauplii were observed to rotate around the pitch axis in the CCW direction only. The direction of rotation around the pitch axis was less ambiguous than the yaw and roll axes because the body remained in focus, with minimal translation during the maneuver. Additionally, the body orientations during the pitch maneuvers could be easily understood. For example, Figure 6b shows that the appendages located on the ventral side of the body are oriented more toward the initial heading direction (Figure 6a). This means that, regardless of whether the ventral or dorsal side is initially in view in Figure 6a, the body rotates around the pitch axis, with the ventral side facing outward. This corresponds to a backflip somersault with CCW rotation around the pitch axis.

Table 2. Clockwise and counterclockwise occurrences of yaw, roll, and pitch rotations.

	Yaw	Roll	Pitch
CW	10	12	0
CCW	8	14	30

4. Discussion

In summary, copepod nauplii exhibited rotations about three axes termed yaw, roll, and pitch. Naupliar stages N1/N2 were observed to perform yaw and roll rotations only, while ≥N3 nauplii were observed to rotate around all three axes. The pitch rotation was observed only in the CCW direction, meaning that ≥N3 nauplii were observed to flip backward, not frontward. Most previous studies have examined the coordinated movement of the paired appendages, providing limited information on body rotation. Our observations are complementary to the studies and provide insight into how nauplii use their appendages to navigate in three-dimensional space.

The nauplius stage is widespread among the crustacea. While the organization of the naupliar body plan constrains movements, comparisons among nauplii from different crustacean groups have shown significant flexibility in locomotion [27]. Specifically, variations in morphology, the involvement of one vs. multiple pairs of appendages, and substantial differences in beat frequencies of the appendage(s) have been reported across taxa [27,36,37]. Copepod nauplii depend on locomotion to survive in an environment where predation risk is high, while food levels can be low. These nauplii are pelagic, capable of high-performance escape swims, and typically start feeding at the N3 stage [15,16,32,38]. The rapid escape swim is produced by the coordinated high-frequency (>100 Hz) beating of all three pairs of appendages (first antenna, second antenna, and mandible) [6,39]. Finding and capturing food is another challenge. For example, on a large scale, the nauplii of *Neocalanus* spp., in the subarctic North Pacific migrate vertically from depth (>200 m) to surface waters (0–50 m) to feed on the spring phytoplankton bloom [40]. At small spatial scales in the millimeter range, feeding behavior includes food-search strategies, as documented by 3D videography, which can be species-specific [14]. At the micrometer scale, the capture of an alga can involve a feeding current, as shown for *Eucalanus pileatus* [13] or more complex turning maneuvers to capture the alga, as shown for *Temora longicornis* and *Acartia tonsa*, using high-speed video [41].

While crustacean nauplii are small (<1 mm), the nauplii of the two target species in this study are among the smallest (length: 0.06 to 0.2 mm). These species operate at a low Reynolds number and require complex maneuvers to change the orientation of their body. Two of the rotations (yaw and roll) were observed in all naupliar stages (N1 to N6). These rotations allow the nauplii to change swim direction in three dimensions and contribute to the typical helical swim patterns described in other studies. Given that all naupliar stages are at risk for predation, the ability to redirect may add to their ability to evade predators. The third rotation, pitch, was only recorded in feeding stages (N3 to N6). The lack of a flexible tail in the N1/N2 may prevent these stages from completing the pitch rotation, which in turn may limit their ability to feed at low Reynolds number effectively. One might speculate that this may have led to the evolution of nauplii that depend on maternal resources to complete two molt cycles before they start to feed.

We conclude with a discussion of the space that is accessible to copepod nauplii and the implications for controlling microscopic robots that propel themselves autonomously. The ability to relocate from one position to another opens the possibility of reaching other positions in space, which depends importantly on the rotational maneuverability of the body. In general, the body does not remain perfectly axisymmetric around the direction of locomotion. Any minor left–right asymmetry in the shape or actuation mechanism prevents the body from following a straight line and instead turns the body to one side, as seen in the yaw rotation. Turning repeatedly to only one side would confine the body to loop around the same circle over time. The ability to yaw freely in either CW or CCW direction is important because it enables the body to escape from the circle and explore any point in the two-dimensional plane perpendicular to the yaw axis. Furthermore, to explore outside this plane, the body must rotate around another axis, e.g., the roll axis. The earliest stages of copepod nauplii display the ability to yaw and roll, and they can be executed in theory by the pronation or supination of a single biramous appendage that splits into two branches, as described here. Despite the apparent inability of the earliest stages of nauplii to rotate directly around the third (pitch) axis, they can explore three-dimensional space in principle. For example, the result corresponding to a

CCW pitch rotation by 90 degrees could be achieved by a time-consuming combination of yaw and roll rotations along a spiral path, e.g., a sequence of 90-degree CCW yaw, CW roll, and CW yaw rotations, in that order. The benefit of growing and developing a flexible tail in the older feeding stages of nauplii is that a similar result to such a sequence can be achieved more efficiently by rotating directly around the pitch axis. The strategies adopted by nauplii are excellent sources of inspiration for designing robots capable of navigating autonomously at microscopic scales. The robots may be configured to respond adequately to external cues, such as sudden changes in light or chemical gradients, with rapid acrobatic maneuvers, as observed in nature.

Supplementary Materials: The following are available online at http://www.mdpi.com/2311-5521/5/2/78/s1. Video S1: Yaw rotation. Video S2: Yaw rotation at higher magnification. Video S3: Roll rotation. Video S4: Roll rotation at higher magnification. Video S5: Pitch rotation. Video S6: Pitch rotation viewed parallel to the pitch axis.

Author Contributions: Conceptualization, K.T.M.N. and D.T.; data curation, K.T.M.N. and K.J.K.; formal analysis, K.T.M.N., K.J.K., L.N.B., P.H.L., and D.T.; writing—review and editing, K.T.M.N., L.N.B., P.H.L., and D.T.; funding acquisition, P.H.L. and D.T. All authors have read and agreed to the published version of the manuscript.

Funding: This research was funded by US National Science Foundation, grant numbers OCE-1235549, to P.H.L. and D. K. Hartline, and CBET-1603929 to D.T., and US Army Research Office, grant number W911NF-17-1-0442 to D.T.

Acknowledgments: The authors would like to thank Curtis Chan and Kyle Nugent for help providing the copepod nauplii, and Rudi Strickler for help setting up the side-view recordings.

Conflicts of Interest: The authors declare no conflict of interest.

References

1. Bergou, A.J.; Ristroph, L.; Guckenheimer, J.; Cohen, I.; Wang, Z.J. Fruit flies modulate passive wing pitching to generate in-flight turns. *Phys. Rev. Lett.* **2010**, *104*, 148101. [CrossRef] [PubMed]
2. Fish, F.E.; Nicastro, A.J.; Weihs, D. Dynamics of the aerial maneuvers of spinner dolphins. *J. Exp. Biol.* **2006**, *209*, 590–598. [CrossRef] [PubMed]
3. Lauga, E. Bacterial hydrodynamics. *Annu. Rev. Fluid Mech.* **2016**, *48*, 105–130. [CrossRef]
4. Goldstein, R.E. Green algae as model organisms for biological fluid dynamics. *Annu. Rev. Fluid Mech.* **2015**, *47*, 343–375. [CrossRef] [PubMed]
5. Gemmell, B.J.; Sheng, J.; Buskey, E.J. Compensatory escape mechanism at low Reynolds number. *Proc. Natl. Acad. Sci. USA* **2013**, *110*, 4661–4666. [CrossRef] [PubMed]
6. Lenz, P.H.; Takagi, D.; Hartline, D.K. Choreographed swimming of copepod nauplii. *J. R. Soc. Interface* **2015**, *12*, 20150776. [CrossRef]
7. Purcell, E.M. Life at low Reynolds number. *Am. J. Phys.* **1977**, *45*, 3–11. [CrossRef]
8. Dreyfus, R.; Baudry, J.; Stone, H.A. Purcell's "rotator": Mechanical rotation at low Reynolds number. *Eur. Phys. J. B* **2005**, *47*, 161–164. [CrossRef]
9. Rizvi, M.S.; Farutin, A.; Misbah, C. Three-bead steering microswimmers. *Phys. Rev. E* **2018**, *97*, 023102. [CrossRef]
10. Jalali, M.A.; Alam, M.R.; Mousavi, S. Versatile low-Reynolds-number swimmer with three-dimensional maneuverability. *Phys. Rev. E* **2014**, *90*, 053006. [CrossRef]
11. Tottori, S.; Zhang, L.; Qiu, F.; Krawczyk, K.K.; Franco-Obregón, A.; Nelson, B.J. Magnetic helical micromachines: Fabrication, controlled swimming, and cargo transport. *Adv. Mater.* **2012**, *24*, 811–816. [CrossRef] [PubMed]
12. Hosney, A.; Klingner, A.; Misra, S.; Khalil, I.S.M. Propulsion and steering of helical magnetic microrobots using two synchronized rotating dipole fields in three-dimensional space. In Proceedings of the 2015 IEEE/RSJ International Conference on Intelligent Robots and Systems (IROS), Hamburg, Germany, 28 September–2 October 2015; pp. 1988–1993.
13. Paffenhöfer, G.A.; Lewis, K.D. Feeding behavior of nauplii of the genus *Eucalanus* (Copepoda, Calanoida). *Mar. Ecol. Prog. Ser.* **1989**, *57*, 129–136. [CrossRef]
14. Paffenhöfer, G.A.; Strickler, J.R.; Lewis, K.D.; Richman, S. Motion behavior of nauplii and early copepodid stages of marine planktonic copepods. *J. Plankton Res.* **1996**, *18*, 1699–1715. [CrossRef]

15. Titelman, J. Swimming and escape behavior of copepod nauplii: Implications for predator-prey interactions among copepods. *Mar. Ecol. Prog. Ser.* **2001**, *213*, 203–213. [CrossRef]
16. Bradley, C.J.; Strickler, J.R.; Buskey, E.J.; Lenz, P.H. Swimming and escape behavior in two species of calanoid copepods from nauplius to adult. *J. Plankton Res.* **2013**, *35*, 49–65. [CrossRef]
17. Eiane, K.; Aksnes, D.L.; Ohman, M.D.; Wood, S.; Martinussen, M.B. Stage-specific mortality of *Calanus* spp. under different predation regimes. *Limnol. Oceanogr.* **2002**, *47*, 636–645. [CrossRef]
18. Turner, J.T. The importance of small planktonic copepods and their roles in pelagic marine food webs. *Zool. Stud.* **2004**, *43*, 255–266.
19. Sampey, A.; McKinnon, A.D.; Meekan, M.G.; McCormick, M.I. Glimpse into guts: Overview of the feeding of larvae of tropical shorefishes. *Mar. Ecol. Prog. Ser.* **2007**, *339*, 243–257. [CrossRef]
20. Svetlichnyy, L. Speed, force and energy expenditure in the movement of copepods. *Oceanology* **1987**, *27*, 497–502.
21. Alcaraz, M.; Strickler, J.R. Locomotion in copepods: Pattern of movements and energetics of *Cyclops*. *Hydrobiologia* **1988**, *167*, 409–414. [CrossRef]
22. Lenz, P.H.; Hartline, D.K. Reaction times and force production during escape behavior of a calanoid copepod, *Undinula vulgaris*. *Mar. Biol.* **1999**, *133*, 249–258. [CrossRef]
23. Buskey, E.J.; Lenz, P.H.; Hartline, D.K. Escape behavior of planktonic copepods in response to hydrodynamic disturbances: High speed video analysis. *Mar. Ecol. Prog. Ser.* **2002**, *235*, 135–146. [CrossRef]
24. Buskey, E.J.; Strickler, J.R.; Bradley, C.J.; Hartline, D.K.; Lenz, P.H. Escapes in copepods: Comparison between myelinate and amyelinate species. *J. Exp. Biol.* **2017**, *220*, 754–758. [CrossRef] [PubMed]
25. Robinson, H.E.; Strickler, J.R.; Henderson, M.J.; Hartline, D.K.; Lenz, P.H. Predation strategies of larval clownfish capturing evasive copepod prey. *Mar. Ecol. Prog. Ser.* **2019**, *614*, 125–146. [CrossRef]
26. Bruno, E.; Andersen Borg, C.M.; Kiørboe, T. Prey detection and prey capture in copepod nauplii. *PLoS ONE* **2012**, *7*, e47906. [CrossRef] [PubMed]
27. Williams, T.A. The nauplius larva of crustaceans: Functional diversity and the phylotypic stage. *Integr. Comp. Biol.* **1994**, *34*, 562–569. [CrossRef]
28. Takagi, D. Swimming with stiff legs at low Reynolds number. *Phys. Rev. E* **2015**, *92*, 023020. [CrossRef] [PubMed]
29. Hayashi, R.; Takagi, D. Metachronal swimming with rigid arms near boundaries. *Fluids* **2020**, *5*, 24. [CrossRef]
30. VanderLugt, K.; Lenz, P. Management of nauplius production in the paracalanid, *Bestiolina similis* (Crustacea: Copepoda): Effects of stocking densities and culture dilution. *Aquaculture* **2008**, *276*, 69–77. [CrossRef]
31. Bathen, K.H. *A Descriptive Study of the Physical Oceanography of Kaneohe Bay, Oahu, Hawaii*; Hawai'i Institute of Marine Biology (Formerly Hawai'i Marine Laboratory): Kaneohe, HI, USA, 1968.
32. Mauchline, J. *The Biology of Calanoid Copepods*; Academic Press: San Diego, CA, USA, 1998.
33. Jennings, H.S. On the significance of the spiral swimming organisms. *Am. Nat.* **1901**, *35*, 369–378. [CrossRef]
34. Crenshaw, H.C. A new look at locomotion in microorganisms: Rotating and translating. *Integr. Comp. Biol.* **1996**, *36*, 608–618. [CrossRef]
35. Becker, L.E.; Koehler, S.A.; Stone, H.A. On self-propulsion of micro-machines at low Reynolds number: Purcell's three-link swimmer. *J. Fluid Mech.* **2003**, *490*, 15–35. [CrossRef]
36. Williams, T.A. A model of rowing propulsion and the ontogeny of locomotion in *Artemia* larvae. *Biol. Bull.* **1994**, *187*, 164–173. [CrossRef] [PubMed]
37. Dahms, H.U.; Fornshell, J.A.; Fornshell, B.J. Key for the identification of crustacean nauplii. *Org. Divers. Evol.* **2006**, *6*, 47–56. [CrossRef]
38. Titelman, J.; Kiørboe, T. Predator avoidance by nauplii. *Mar. Ecol. Prog. Ser.* **2003**, *247*, 134–149.
39. Wadhwa, N.; Andersen, A.; Kiørboe, T. Hydrodynamics and energetics of jumping copepod nauplii and copepodids. *J. Exp. Biol* **2014**, *217*, 3085–3094.

40. Mackas, D.L.; Tsuda, A. Mesozooplankton in the eastern and western subarctic Pacific: Community structure, seasonal life histories, and interannual variability. *Prog. Oceangr.* **1999**, *43*, 335–363. [CrossRef]
41. Borg, M.A.; Bruno, E.; Kiørboe, T. The kinematics of swimming and relocation jumps in copepod nauplii. *PLoS ONE* **2012**, *7*, e47486. [CrossRef]

© 2020 by the authors. Licensee MDPI, Basel, Switzerland. This article is an open access article distributed under the terms and conditions of the Creative Commons Attribution (CC BY) license (http://creativecommons.org/licenses/by/4.0/).

Article

Jellyfish and Fish Solve the Challenges of Turning Dynamics Similarly to Achieve High Maneuverability

John O. Dabiri [1], Sean P. Colin [2,3], Brad J. Gemmell [4], Kelsey N. Lucas [5], Megan C. Leftwich [6] and John H. Costello [3,7,*]

1. Graduate Aerospace Laboratories and Mechanical Engineering, California Institute of Technology, Pasadena, CA 91125, USA; jodabiri@caltech.edu
2. Marine Biology and Environmental Science, Roger Williams University, Bristol, RI 02809, USA; scolin@rwu.edu
3. Whitman Center, Marine Biological Laboratory, Woods Hole, MA 02543, USA
4. Department of Integrative Biology, University of South Florida, Tampa, FL 33620, USA; bgemmell@usf.edu
5. School for Environment and Sustainability, University of Michigan, Ann Arbor, MI 48109, USA; kelsey.n.lucas@gmail.com
6. Department of Mechanical and Aerospace Engineering, The George Washington University, Washington, DC 20052, USA; megan.leftwich@gmail.com
7. Biology Department, Providence College, Providence, RI 02918, USA
* Correspondence: costello@providence.edu

Received: 21 March 2020; Accepted: 28 June 2020; Published: 30 June 2020

Abstract: Turning maneuvers by aquatic animals are essential for fundamental life functions such as finding food or mates while avoiding predation. However, turning requires resolution of a fundamental dilemma based in rotational mechanics: the force powering a turn (torque) is favored by an expanded body configuration that maximizes lever arm length, yet minimizing the resistance to a turn (the moment of inertia) is favored by a contracted body configuration. How do animals balance these opposing demands? Here, we directly measure instantaneous forces along the bodies of two animal models—the radially symmetric *Aurelia aurita* jellyfish, and the bilaterally symmetric *Danio rerio* zebrafish—to evaluate their turning dynamics. Both began turns with a small, rapid shift in body kinematics that preceded major axial rotation. Although small in absolute magnitude, the high fluid accelerations achieved by these initial motions generated powerful pressure gradients that maximized torque at the start of a turn. This pattern allows these animals to initially maximize torque production before major body curvature changes. Both animals then subsequently minimized the moment of inertia, and hence resistance to axial rotation, by body bending. This sequential solution provides insight into the advantages of re-arranging mass by bending during routine swimming turns.

Keywords: propulsion; rotational physics; convergent evolution; torque; moment of inertia; animal movement

1. Introduction

The study of aquatic locomotion has primarily focused on the dynamics and energetics of linear, unidirectional swimming. This approach has yielded important insights but largely reflects longstanding constraints in the empirical measurement, numerical simulation, and theoretical modeling of animal swimming. Experiments conducted in a water channel constrain animal swimming to the single direction of the oncoming flow. With the exception of notable efforts to quantify C-start and S-start behaviors of some fishes [1–3], experimental [4,5] and theoretical [6] biomechanical models of animal swimming focus primarily on linear translation. The implicit assumption that swimming is primarily unidirectional has influenced prevailing notions regarding the kinematic parameters that are most important for efficient swimming and body design. Specifically, the observation that swimming animals maintain nearly constant values of Strouhal number $St = fA/U$ (where f is the stroke frequency, A is the stroke amplitude, and U is the unidirectional, steady state swimming speed) has encouraged many efforts to explain the efficiency of animal swimming on the basis of the unidirectional swimming parameters that define the Strouhal number [7–9]. Likewise, other measures employed to compare swimming efficiency between animals, such as cost of transport [10,11] and Froude efficiency [12–14] inherently place animal swimming within the context of linear pathways between points in a fluid.

This emphasis on unidirectional swimming belies the fact that actual animal swimming in nature is rarely linear, but instead, is more typically characterized by frequent changes in direction that are mediated by turning maneuvers. The importance of turning has long been documented in studies of aquatic animal ecology. Efforts to model the circuitous trajectories of animals have often focused on Brownian motion or Levy walks [15–17]. Regardless of behavioral assumptions about swimmers, many studies of empirically measured pathways have demonstrated that across a variety of spatial scales, swimming animals exhibit predominantly non-linear pathways with frequent turns that change their trajectories. Recognition that swimmers in nature turn frequently is important from a biomechanical perspective because turning maneuvers require rotational motions of the swimmer's major body axis. The mechanics of rotational motion parallel, but differ from, the more studied mechanics of linear translation by swimmers (Figure 1). In contrast to the large body of knowledge concerning thrust production and force generation during linear swimming, there is not a similar body of mechanical information evaluating torque generation and moment of inertia minimization by flexible bodies such as animal swimmers. Consequently, greater understanding of maneuverability by animal swimmers requires deeper examination of their rotational mechanics to complement existing knowledge of their translational mechanics.

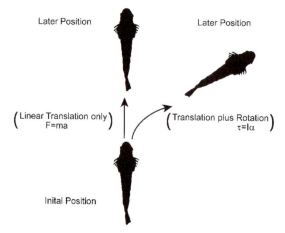

Figure 1. Swimming turns require both translational and rotational components of motion. The mechanics of these components are described by parallel but different physical terms for translational force and rotational torque (F = thrust force, m = mass, a = acceleration; τ = torque, I = moment of inertia and α = rotational acceleration).

Evaluation of rotational mechanics involves a previously unaddressed issue that is essential for turning by animal swimmers. The same body configurations that maximize the forces powering a turn (torque) also maximize that body's the resistance to turning (moment of inertia). Torque (τ) generation relies upon a force (F) applied at a distance (r) from the axis of rotation (r is also termed the lever arm) according to the relationship $\tau = Fr$. The longer the lever arm, r, the greater is the torque applied by a limited force to power a turn. Consequently, the most force-efficient body configuration for turning is an elongate or expanded body that maximizes r and requires the least amount of force to affect axial rotation. However, there is an inherent problem with expanded body forms for turning because expanded bodies also maximize the moment of inertia (I), that resists angular rotation of a body according to the relationship $\tau = I\alpha$ where α is angular acceleration. For a limited torque, the greatest angular acceleration will be achieved when the body's moment of inertia (I) is minimized. I depends upon the arrangement of a body's mass around the axis of rotation according to the relationship $I_P = \sum_{i=1}^{N} m_i r_i^2$ where I_P represents the sum moments of inertia for the constituent parts ($i \ldots N$) of a swimmers body with m_i denoting that body part's mass (e.g., the head or tail of the body) and r_i its distance from the whole-body center of rotation. There are straightforward means to minimize I_P, e.g., the mass of the body can be re-arranged to place body components closer to the axis of whole-body rotation. This is commonly achieved by bending body parts closer to the axis during a turn. Flexible bodies that allow bending by animal swimmers permit dramatically greater angular velocities during turns than are possible for rigid animal bodies or rigid human-engineered structures [18]. However, it remains unclear how these flexible swimmers resolve the fundamentally conflicting demands of high torque production (expanded body configuration) with those of low moment of inertia (contracted body configuration) to achieve high turning performance. The results are important for understanding maneuverability by swimming animals, and potentially, human engineered vehicles.

We hypothesized that the high frequency and energetic demands of turning by natural swimmers could produce a selective force on swimming performance that might lead to similar solutions for widely divergent animal models. Such patterns would be missed by the conventional biomechanical focus on unidirectional translational swimming, yet are essential for efficient aquatic locomotion by these swimmers in their natural environments.

To evaluate this question broadly, we used two model species with extremely divergent body types, neural organization, and phylogenetic relatedness. The jellyfish *Aurelia aurita* is a member of the oldest animal group to use muscle-driven swimming and one of the most energetically efficient metazoan swimmers [11]. Medusae such as *A. aurita* are characterized by a radially symmetric body plan with a comparatively simple level of neuromuscular organization [19]. By contrast, the zebrafish *Danio rerio* represents the evolution of a bilaterally symmetric body plan with comparatively complex neuromuscular organization representative of modern fish species [20]. In both cases, we quantified their natural swimming motions using a combination of high-speed videography and laser-based flow measurements.

2. Materials and Methods

2.1. Animals and Imaging

The zebrafish (*Danio rerio*) used in this study were adults acquired from the Zebrafish Facility at the Marine Biological Laboratory (MBL). All procedures were in accordance with standards set by the National Institutes of Health and approved by the Institutional Animal Care and Use Committee at the MBL. Zebrafish were maintained at room temperature (23–25 °C) in 37 L aquaria until imaged while swimming. Swimming and turning behaviors were recorded as individual fish swam along the center of an acrylic raceway tank (1.5 × 0.5 m). *Aurelia aurita* medusae were obtained from the New England Aquarium and maintained at 25 °C in 20 l aquaria. Medusae were recorded while freely swimming in a 0.3 × 0.1 × 0.25 m glass vessel, using methods reported previously [11]. Many individuals of both species were recorded, but only those that swam within the laser light plane could be used for analysis. The number of separate individuals satisfying this criterion was greatest for the start of the turn (n = 10 for both species). A number of individuals subsequently moved out of the laser sheet while completing a turn. Time course analysis for full turns was limited to separate individuals that completed full turns within the laser light sheet (n = 4 for zebrafish and n = 6 for medusae).

2.2. Particle Image Velocimetry (PIV)

We used high-speed digital particle image velocimetry (PIV) to obtain resulting flow fields around the fish and medusae. Recordings were acquired by a high-speed digital video camera (Fastcam 1024 PCI; Photron, San Diego, CA, USA) at 1000 frames per second and at a spatial resolution of 1024 × 1024 pixels with a scale factor of 0.178 mm per pixel. Seeding particles (10 µm hollow glass beads; Potters Industries, Malvern, PA, USA) were laser-sheet illuminated for PIV measurements. Medusae were illuminated with a laser sheet (680 nm, 2W continuous wave; LaVision, Ypsilanti, MI, USA) oriented perpendicular to the camera's optical axis to provide a distinctive body outline for image analysis and to ensure the animal remained in-plane, which ensures accuracy of 2D estimates of position and velocity. The semitransparent bodies of medusae allowed a single laser light sheet passing through the central axis of the body to illuminate fluid surrounding the entire body. Fish were not transparent and so were illuminated by two laser sheets (532 nm, 600 mW continuous wave, Laserglow Technologies, North York, ON, Canada) mounted in the same plane on opposite sides of the tank to eliminate shadows on either side of the body as each animal swam within the field of view [21].

Fluid velocity vectors for both fish and medusae were determined from sequential images using a cross-correlation algorithm (LaVision software). Image pairs were analyzed with shifting overlapping interrogation windows of a decreasing size of 32 × 32 pixels to 16 × 16 pixels. Masking of the body of the fish before image interrogation confirmed the absence of surface artifacts in the PIV measurements. While the medusae were not masked for velocity analyses, our previous work with medusae [11], indicated that adverse effects from surface artifacts are minimal.

2.3. Pressure and Torque Measurement

Direct measurements of instantaneous forces acting along animal bodies were made throughout complete turning sequences. These measurements were produced at high spatial and temporal resolution, providing instantaneous values at highly localized points on the body [22–24], contrasting with, for example, net force calculation based on vortex circulation. Our approach involved converting velocity fields collected via PIV through a custom program in MATLAB that computed the corresponding pressure fields. The algorithm integrates the Navier–Stokes equations along eight paths emanating from each point in the field of view and terminating at the boundaries of the field of view. The pressure at each point is determined by computing the median pressure from the eight integration results. Bodies of the fish and medusae were masked prior to computation to prevent surface artefacts in the pressure and torque results. Masks were generated using a custom MATLAB (Mathworks, Inc., Natick, MA, USA) program that automatically identified the boundary of the animal body based on image contrast at the interface between the animal body and the surrounding fluid, and body outlines were smoothed prior to later analyses. The zebrafish's anatomy allowed for these outlines to enclose the body while allowing flow calculations directly alongside all surfaces, but for medusae, the semitransparent bell and opaque gonads outside the laser light sheet interfere with the view of the subumbrellar surface within the bell cavity. As such, the outlines around the medusae traced the exumbrellar surface and, on the oral side of the body, the bell margin to prevent erroneous pressure calculations within the bell cavity from affecting pressure calculations in the areas adjacent to the medusan oral sides. These methods have been previously validated against experimental and computational data, including numerical simulations of anguilliform swimming [22] and direct force and torque measurements of a flapping foil [24]. The MATLAB code is available for free download at http://dabirilab.com/software.

The fluid force normal to the body surface due to the local fluid pressure was determined by integrating the calculated pressure along the corresponding surfaces of the body [24]. Validations against measurements made on physical models show that these calculation techniques based on 2D PIV images are robust to a small degree of out-of-plane flow such as that induced by a fish's slight rolling motions during turns, so long as the fish remains centered in the imaging plane [24]. The body outline of each animal was divided into segments of equal length (zebrafish: 84 segments, medusa: 70–85 segments) for spatial integration. Again, for medusae, bell components outside the laser sheet can interfere with images of the subumbrellar surface within the bell cavity, so surface segments on the oral side of the body traced the bell margin even as it protruded out of the PIV imaging plane. Although these bell margin surface segments were required to mask the animal body during pressure calculation as indicated above, the central segments—defined as the central two-thirds of the bell's radius—did not represent surfaces visible within the PIV laser light sheet, and so forces and torques calculated on these central bell segments were not included in later analyses. The areas where calculations were conducted are visible in the force vector plots in Figure 2i–l.

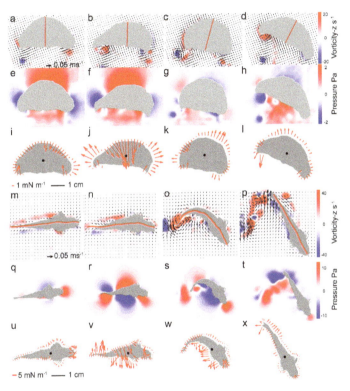

Figure 2. Turning kinematics and fluid pressure for representative medusa (*Aurelia aurita*, 30° rotation, profiled in Figure A1d in Appendix A) and zebrafish (*Danio rerio*, 62° rotation, profiled in Figure A2c) turns. The red line shows the midline of the medusa (**a–d**) and the fish (**m–p**) throughout the turn, along with PIV vector and vorticity fields. Pressure fields around the medusa (**e–h**) and the fish (**q–t**) demonstrate that both animals generate large, asymmetric pressure gradients around their bodies (panels (**f**) and (**r**), respectively) before major body orientation shifts (illustrated by the midline position). Force vectors exerted on the animal due to local fluid pressure at the medusa (**i–l**) and zebrafish (**u–x**) body surface indicated in red arrows. Note that force vectors, and hence torques, were not calculated on the central region of the oral surface of the jellyfish (the bottom of the bell), as the bell margin in this region protrudes outward from the 2D imaging plane and blocks the view of the subumbrellar surface within the bell cavity, the surface where forces and torques would actually act. Black circles represent the center of mass in each of the latter panels. Note that during peak torque periods, forces along the body stabilize the center of mass while causing rotation of extended body regions such as the bell margin of medusae (**j**) and caudal fin of fish (**v**). For jellyfish, the most rapid rotation occurs during bell contraction and bell relaxation may be accompanied by negative torque (**l**) that brakes bell rotation.

Because the surface geometry was specified in a single plane, the force calculations were evaluated per unit depth (i.e., giving units of Newtons per meter of depth perpendicular to the measurement plane). The corresponding torque was calculated as the vector product of the moment arm from each location on the body surface to the center of mass, and the local force due to pressure at the same location on the body surface. The resulting torque calculations have units of Newton-meters per meter, corresponding to the aforementioned planar measurements. MATLAB codes for force and torque calculations similar to those conducted presently as well as the segment-making methods have been validated in earlier work [24] and are available on Github (https://github.com/kelseynlucas).

2.4. Turning Equations of Motion

The mass moment of inertia of a body is a measure of how its mass is distributed relative to a reference axis, often taken as the geometric centroid. It is given by

$$I = \int_V r^2 dm \tag{1}$$

where V is the region occupied by the body mass, and r is the distance of each infinitesimal portion of body mass from the reference axis. In the present case, this mass moment of inertia was approximated using the area moment of inertia, which is a measure of how the body area in a cross section is distributed relative to the reference axis:

$$I_A = \iint_A r^2 dA \tag{2}$$

where A is the region occupied by a two-dimensional cross-section of the body. The cross-section in the present measurements was the body symmetry plane illuminated by the laser sheet during PIV measurements. The area moment of inertia (henceforth called the moment of inertia for brevity) was calculated using a custom program in MATLAB as described in the following section.

The torque exerted on a body is related to changes to both its angular motion and its moment of inertia by the following relation:

$$\tau = \frac{d(I\omega)}{dt} = I\frac{d\omega}{dt} + \omega\frac{dI}{dt} \tag{3}$$

where ω is the angular velocity of the body. The first term of the summation incorporates the rate of change of angular velocity, i.e., the angular acceleration. The second term depends on the change in the moment of inertia, i.e., changes in body shape or mass.

2.5. Moment of Inertia and Angular Velocity Measurements

Calculations of the moment of inertia for turning sequences used the same smoothed animal body outlines automatically detected for pressure and torque calculation. A separate custom MATLAB algorithm subsequently calculated the moment of area for each image. Animal bodies were partitioned as for the force and torque measurements above, with each of segment of area a_i having a centroid located at distance r_i from the whole body centroid. The area moment of inertia for each frame p was then calculated as:

$$I_P \approx \sum_{i=1}^{N} a_i r_i^2 \tag{4}$$

where the summation was taken over the N body segments. Angular velocities of zebrafish during turns used local body surface position changes to calculate the angle of the line segment connecting the anterior head region with that of the body centroid. The rate of change of that angle in a lab-fixed frame determined the fish angular velocity. The hemi-ellipsoidal shape of medusae and shifts within the bell during contraction required a different approach for angular measurements. Medusan angular changes were measured by changes of relatively fixed structures within the bell, the gonads, during medusan turning. The angle of the selected gonads were measured relative to the lab-fixed frame in successive images using Image J v1.48 software (National Institutes of Health, Bethesda, MD, USA).

3. Results

Jellyfish (Figure 2a–l)-and zebrafish (Figure 2m–x) both exhibited frequent bouts of turning, during which flow measurements revealed pronounced changes in fluid velocities and pressure fields in the water adjacent to the animal (Figure 2f,r, for jellyfish and fish, respectively). These substantial pressure fields preceded the more pronounced body motions that occurred during the subsequent turn that changed the animal swimming direction (Figure 2c,o, respectively).

Examination of the body shape during the period of transient pressure buildup led to the discovery of a small, rapid asymmetric shift in the curvature of the animal body immediately preceding the turn for both the jellyfish (1.5 ± 1.0 percent change in curvature, n = 10 individuals) and the zebrafish (0.8 ± 0.2 percent, n = 10 individuals). Although the amplitude of this initial body bend was small, it occurred over a sufficiently short period of time—few milliseconds—that the corresponding acceleration of the body was large relative to accelerations during unidirectional swimming. These fluid accelerations occurred along much of the body surface as the extreme outset of the turn (Figure 3). The measured peak accelerations preceding the turn were over 1 m s^{-2}. This motion was transmitted to the adjacent water via a process known as the acceleration reaction or added-mass effect [25].

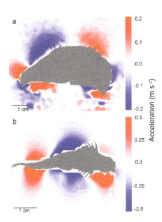

Figure 3. Rapid fluid accelerations during turn initiation give rise to high torque forces along the bodies of jellyfish and fish. Fluid acceleration (positive values correspond to vertical motion toward bottom of page) along animal bodies during turn initiation by medusa (**a**) *Aurelia aurita* and zebrafish (**b**) *Danio rerio*. Fluid accelerations in both panels are for the same turning sequences as depicted in Figure 2, so that the acceleration field in panel (**a**) corresponds to the high pressure state of Figure 2f, while panel (**b**) corresponds to that of Figure 2r.

Because the water is effectively incompressible, the fluid in contact with the body responded to the high local body acceleration by an increase in the local fluid pressure where the body was advancing (pushing the water), and a decrease in the local pressure where the body surface retreated from the local water (pulling the water with it). When integrated over the full animal body, the pressure field created by the small, asymmetric body bending leads to a large net torque capable of turning the organism toward a new heading. The more pronounced body motions that occur after the generation of this pressure field do not contribute greatly to torque generation, but they do reduce the moment of inertia of the body (Figure 4; see also Figures A1 and A2). Therefore, the body kinematics that follow peak pressure generation enhance the effect of the generated torque by amplifying the resulting angular acceleration so that the body rotates rapidly through a turn. This sequence of asymmetric body kinematics that initially maximizes torque forces and subsequently minimizes the moment of inertia resolves the fundamental competition between these two components of rotational motion during turns. Although the maximum torque generation and minimum moment of inertia do not occur simultaneously (Figure 5), the inertia of the fluid and of the animal body allows the initial pressure transient to affect subsequent turning dynamics even as fluid viscosity resists body acceleration during a turn.

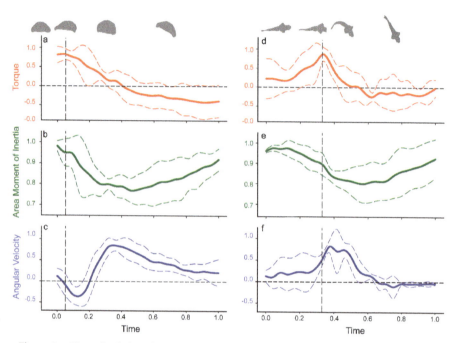

Figure 4. Normalized data for comparison of turning variables between jellyfish and fish. Patterns represent data for replicate individuals during variable turn excursions (medusa *Aurelia aurita*, panels (**a–c**), n = 6; bell diameters 1.8–5.4 cm, range in turn angles 13–53°; zebrafish *Danio rerio*, panels (**d–f**), n = 4, fish lengths 3.2–4.4 cm, range in turn angles 17–95°). Data for each replicate turn was divided into a uniform number of sample intervals and each variable (time, area moment of inertia, angular velocity and torque) was normalized by the highest value of each replicate sequence so that all variables could be expressed in dimensionless form with a maximum value of 1. Solid curves represent the mean value and dashed lines represent one standard deviation above or below the mean for each sample interval. Note that peak values do not always reach 1 because they are averages of all the turns and not all the peak values occurred in the same time interval for every turn. The original, non-normalized data for each individual replicate are displayed in Figure A1 (medusa) and Figure A2 (zebrafish).

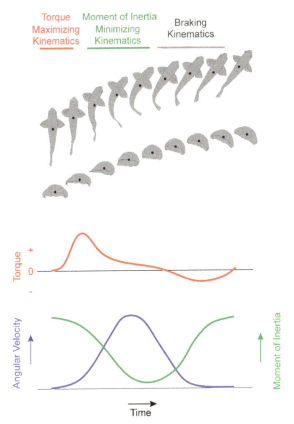

Figure 5. Conceptual summary of turning dynamics by the medusa (*Aurelia aurita*) and the zebrafish (*Danio rerio*). Arrows for each axis represent increasing magnitude for that variable. A turn is initiated by a subtle body bend, which builds torque before the animal turns (changes heading). After peak torque production, the animal bends its body more radically to minimize its moment of inertia. This decreases the body's resistance to rotational motion while increasing angular velocity and turning the animal. The turning sequence ends as negative torque brakes the turning rotation when the body returns to its extended configuration with high moment of inertia and low angular velocity. Black circles represent the center of mass for each body image.

4. Discussion

We observed strikingly similar turning dynamics for both the jellyfish and the zebrafish, despite their substantially different body organization and swimming kinematics (Figures 2–4). The dynamical importance of the observed pressure fields for both the jellyfish and zebrafish was confirmed by computing the net torque (per meter depth) and area moment of inertia of the body. For turns of varying net change in heading, the initial pressure pattern created by the animals was nearly constant. The ultimate magnitude of each turning maneuver was instead modulated by asymmetrical changes in body shape that tuned the moment of inertia and thereby controlled the angular acceleration of the body. In all cases, the relationships between pressure measurements and turning kinematics followed a similar sequential pattern (Figure 4).

An essential feature of animal turning by the mechanisms described here is the flexibility of the body, which enables the animal to dynamically redistribute its mass to manipulate the lever arm of the propulsive surfaces used to initiate the turn (e.g., the bell margin of the jellyfish and the caudal fin of the zebrafish) and the body moment of inertia (Figure 5). For animal swimmers with flexibility and size scales favoring this process, the performance advantages of this turning strategy may select for very similar turning kinematics despite the vastly different animal forms studied here.

While the present results motivate further study of turning in other swimming animals whose locomotion lies between jellyfish and zebrafish, we anticipate that extension of these findings will depend upon scaling factors that influence the size range over which this approach is effective. In the regime of swimming at low Reynolds numbers (Re = ULv^{-1}, where U and L are the nominal animal swimming speed and size, respectively, and v is the kinematic viscosity of the water), angular momentum generated during periods of maximum torque would experience rapid viscous dissipation, leaving little remaining angular momentum to complete the turn during the subsequent period of major body bending. For large animals with body lengths on the order of tens of meters, power requirements for rapid body bending may exceed the available muscle capacity. In geometrically similar animals, angular acceleration scales to the −2/3 power of body mass [26], making it more difficult for large animals to generate the initial pressure transient or to alter their moment of inertia through body rearrangement to increase their angular velocity. Hence, very large swimmers such as whales may not bend as readily as smaller animal swimmers such as zebrafish [27]. However, the majority of animal swimmers exist within the millimeter to meter size range [28,29], in which a time-varying lever arm enabled by body bending would provide favorable performance advantages relative to rigid body turning mechanics.

Although the patterns that we describe here—torque maximization followed by concomitant alterations in moment of inertia and angular velocity—may appear unexpected for animal swimmers, a well-developed body of research in the field of human gymnastic diving provides a more intuitive guide to the mechanics of animal turning. Human divers generate all of their angular momentum before they leave the springboard and subsequent alterations in turning velocity come about solely by modulation of the diver's moment of inertia [30]. The diver's ability to rapidly rotate through somersaults and aerial maneuvers depends on the ability to redistribute body mass and alter the diver's moment of inertia [31]. Although the animal models documented here capitalized on self-generated pressure fields rather than a springboard, they utilized analogous patterns to human divers for increasing angular velocity by decreasing moment of inertia during through turning maneuvers. The essential physical relationships between time-varying lever arm deployment, moment of inertia and angular velocity provide a very basic mechanical process enabling rapid turning. For animal swimmers within the size scales favoring this process, the performance advantages of this sequence may select for very similar turning kinematics and provide insight into the convergence of very different animal forms, such as medusae and fish, on similar turning mechanics.

These observations of a large dynamical impact from small kinematic shifts can motivate further study of the neuromuscular control of aquatic locomotion and engineered systems that aim to be inspired by animal swimming. In particular, while nature has not converged upon unidirectional locomotion that leverages similar kinematic subtleties as in turning (i.e., steady, straight swimming does not exhibit the small body motions observed here), it might be feasible to achieve net propulsion using such an approach in a robotic system. The pronounced pressure fields observed presently in the jellyfish and zebrafish are incompatible with unidirectional translation, as they achieve high net torque but low net force due to the balance of high and low pressure on either side of the animal. However, it is conceivable that modified kinematics could result in net propulsive force.

More broadly, an appreciation of the important role of turning maneuvers in the success of aquatic locomotion can influence efforts to understand the role of physical forces in the evolution and ecology of other animal swimmers. The methods employed here to study freely swimming organisms and to quantify their dynamics in terms of pressure field manipulations provide a powerful tool to enable new insights into aquatic locomotion. The solution arrived at by our study organisms allows them to initially maximize torque production before major body curvature changes that subsequently minimize the moment of inertia by bending. Further testing with other animal swimmers will be important for evaluating whether this pattern has influenced the widespread capability of swimmers to re-arrange their mass by flexible bending. Application of similar non-invasive approaches can provide new pathways to understanding the complex physical exchanges that take place between animals and their surrounding fluids.

Author Contributions: All authors conceived the research; S.P.C., B.J.G., M.C.L., and J.H.C. collected animal measurements; all authors analyzed data; J.O.D. and J.H.C. wrote initial manuscript; all authors contributed to revisions. All authors have read and agreed to the published version of the manuscript.

Funding: Funding for this work was provided by the US National Science Foundation (1511333 to J.O.D., 1510929 to S.P.C., 1511996 to B.J.G., 1511721 to J.H.C.) and the Office of Naval Research (000141712248 to M.C.L., N00140810654 to J.H.C.). K.N.L. was supported by a National Science Foundation Graduate Research Fellowship under grant DGE-1745303.

Acknowledgments: We thank Steve Spina and Chris Doller of the New England Aquarium for providing *A. aurita* and Jonathan Gitlin of the Marine Biological Laboratory for providing *Danio rerio* used in our experimental work.

Conflicts of Interest: The authors declare no conflict of interest.

Appendix A

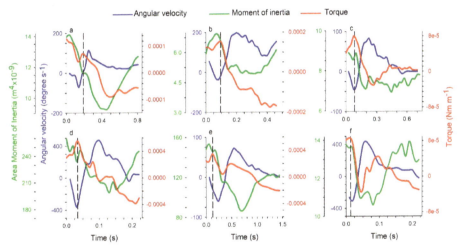

Figure A1. Turning parameters for replicate medusae (*Aurelia aurita*) executing turns of different magnitude. Variable designations are same as in Figure 1: torque per unit depth (red line), angular velocity (blue line) and moment of inertia (green line). Bell diameter and total turn angle for each turn: (**a**) 2.7 cm, 53°, (**b**) 1.8 cm, 50°, (**c**) 2.3 cm, 13°, (**d**) 4.9 cm, 30°, (**e**) 5.4 cm, 20°, (**f**) 2.5 cm, 23°. Local peak in torque is indicated by vertical dashed line each panel.

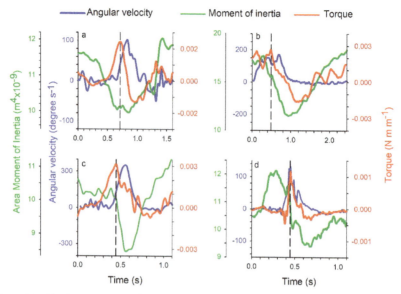

Figure A2. Turning parameters for replicate zebrafish (*Danio rerio*) executing turns of different magnitude. Variable designations are same as in Figure 1: torque per unit depth (red line), angular velocity (blue line) and moment of inertia (green line). Fish body standard length and total turn angle for each turn: (**a**) 4.4 cm, 17°, (**b**) 3.5 cm, 95°, (**c**) 3.2 cm, 62°, (**d**) 3.3 cm, 24°. Local peak in torque is indicated by vertical dashed line each panel. The fluctuations in torque near the end of the turn cycle in panels A–C are within 0.001 mN/m^2 and are within the margin of error that includes zero (Figure 2d).

References

1. Domenici, P.; Blake, R. The kinematics and performance of fish fast-start swimming. *J. Exp. Biol.* **1997**, *200*, 1165–1178.
2. Schriefer, J.E.; Hale, M.E. Strikes and startles of northern pike (*Esox lucius*): A comparison of muscle activity and kinematics between S-start behaviors. *J. Exp. Biol.* **2004**, *207*, 535–544. [CrossRef] [PubMed]
3. Domenici, P. Context-dependent variability in the components of fish escape response: Integrating locomotor performance and behavior. *J. Exp. Zool. Part A Ecol. Genet. Physiol.* **2010**, *313*, 59–79. [CrossRef] [PubMed]
4. Quinn, D.B.; Lauder, G.V.; Smits, A.J. Scaling the propulsive performance of heaving flexible panels. *J. Fluid Mech.* **2014**, *738*, 250–267. [CrossRef]
5. Triantafyllou, M.S.; Techet, A.H.; Hover, F.S. Review of experimental work in biomimetic foils. *IEEE J. Ocean. Eng.* **2004**, *29*, 585–594. [CrossRef]
6. Alben, S. Simulating the dynamics of flexible bodies and vortex sheets. *J. Comput. Phys.* **2009**, *228*, 2587–2603. [CrossRef]
7. Triantafyllou, M.; Triantafyllou, G.; Gopalkrishnan, R. Wake mechanics for thrust generation in oscillating foils. *Phys. Fluids A Fluid Dyn.* **1991**, *3*, 2835–2837. [CrossRef]
8. Taylor, G.K.; Nudds, R.L.; Thomas, A.L. Flying and swimming animals cruise at a Strouhal number tuned for high power efficiency. *Nature* **2003**, *425*, 707. [CrossRef]
9. Eloy, C. Optimal Strouhal number for swimming animals. *J. Fluids Struct.* **2012**, *30*, 205–218. [CrossRef]
10. Schmidt-Nielsen, K. Locomotion: Energy cost of swimming, flying, and running. *Science* **1972**, *177*, 222–228. [CrossRef]
11. Gemmell, B.J.; Costello, J.H.; Colin, S.P.; Stewart, C.J.; Dabiri, J.O.; Tafti, D.; Priya, S. Passive energy recapture in jellyfish contributes to propulsive advantage over other metazoans. *Proc. Natl. Acad. Sci. USA* **2013**, *110*, 17904–17909. [CrossRef] [PubMed]
12. Lighthill, M. Note on the swimming of slender fish. *J. Fluid Mech.* **1960**, *9*, 305–317. [CrossRef]

13. Vogel, S. *Life in Moving Fluids: The Physical Biology of Flow*; Princeton University Press: Princeton, NJ, USA, 1996.
14. Sfakiotakis, M.; Lane, D.M.; Davies, J.B.C. Review of fish swimming modes for aquatic locomotion. *IEEE J. Ocean. Eng.* **1999**, *24*, 237–252. [CrossRef]
15. Viswanathan, G.M.; Buldyrev, S.V.; Havlin, S.; Da Luz, M.; Raposo, E.; Stanley, H.E. Optimizing the success of random searches. *Nature* **1999**, *401*, 911. [CrossRef] [PubMed]
16. Humphries, N.E.; Queiroz, N.; Dyer, J.R.; Pade, N.G.; Musyl, M.K.; Schaefer, K.M.; Fuller, D.W.; Brunnschweiler, J.M.; Doyle, T.K.; Houghton, J.D. Environmental context explains Lévy and Brownian movement patterns of marine predators. *Nature* **2010**, *465*, 1066. [CrossRef]
17. Reynolds, A.M. Current status and future directions of Lévy walk research. *Biol. Open* **2018**, *7*, bio030106. [CrossRef]
18. Fish, F.E.; Kolpas, A.; Crossett, A.; Dudas, M.A.; Moored, K.W.; Bart-Smith, H. Kinematics of swimming of the manta ray: Three-dimensional analysis of open water maneuverability. *J. Exp. Biol.* **2018**, *221*, jeb166041. [CrossRef]
19. Costello, J.H.; Colin, S.P.; Dabiri, J.O. Medusan morphospace: Phylogenetic constraints, biomechanical solutions, and ecological consequences. *Invertebr. Biol.* **2008**, *127*, 265–290. [CrossRef]
20. Severi, K.E.; Portugues, R.; Marques, J.C.; O'Malley, D.M.; Orger, M.B.; Engert, F. Neural control and modulation of swimming speed in the larval zebrafish. *Neuron* **2014**, *83*, 692–707. [CrossRef]
21. Gemmell, B.J.; Fogerson, S.M.; Costello, J.H.; Morgan, J.R.; Dabiri, J.O.; Colin, S.P. How the bending kinematics of swimming lampreys build negative pressure fields for suction thrust. *J. Exp. Biol.* **2016**, *219*, 3884–3895. [CrossRef]
22. Dabiri, J.O.; Bose, S.; Gemmell, B.J.; Colin, S.P.; Costello, J.H. An algorithm to estimate unsteady and quasi-steady pressure fields from velocity field measurements. *J. Exp. Biol.* **2014**, *217*, jeb092767. [CrossRef]
23. Gemmell, B.J.; Colin, S.P.; Costello, J.H.; Dabiri, J.O. Suction-based propulsion as a basis for efficient animal swimming. *Nat. Commun.* **2015**, *6*, 8790. [CrossRef] [PubMed]
24. Lucas, K.N.; Dabiri, J.O.; Lauder, G.V. A pressure-based force and torque prediction technique for the study of fish-like swimming. *PLoS ONE* **2017**, *12*, e0189225. [CrossRef] [PubMed]
25. Daniel, T.L. Unsteady aspects of aquatic locomotion. *Am. Zool.* **1984**, *24*, 121–134. [CrossRef]
26. Carrier, D.R.; Walter, R.M.; Lee, D.V. Influence of rotational inertia on turning performance of theropod dinosaurs: Clues from humans with increased rotational inertia. *J. Exp. Biol.* **2001**, *204*, 3917–3926. [PubMed]
27. Domenici, P. The scaling of locomotor performance in predator–prey encounters: From fish to killer whales. *Comp. Biochem. Physiol. Part A Mol. Integr. Physiol.* **2001**, *131*, 169–182. [CrossRef]
28. Friedman, C.; Leftwich, M. The kinematics of the California sea lion foreflipper during forward swimming. *Bioinspir. Biomim.* **2014**, *9*, 046010. [CrossRef]
29. Nesteruk, I.; Passoni, G.; Redaelli, A. Shape of aquatic animals and their swimming efficiency. *J. Mar. Biol.* **2014**, *2014*, 470715. [CrossRef]
30. McCaw, S. *Biomechanics for Dummies*; John Wiley & Sons: Hoboken, NJ, USA, 2014.
31. Frohlich, C. The physics of somersaulting and twisting. *Sci. Am.* **1980**, *242*, 154–165. [CrossRef]

© 2020 by the authors. Licensee MDPI, Basel, Switzerland. This article is an open access article distributed under the terms and conditions of the Creative Commons Attribution (CC BY) license (http://creativecommons.org/licenses/by/4.0/).

Article

An Elastic Collision Model for Impulsive Jumping by Small Planktonic Organisms

Houshuo Jiang

Applied Ocean Physics & Engineering Department, Woods Hole Oceanographic Institution, Woods Hole, MA 02543, USA; hsjiang@whoi.edu

Received: 29 June 2020; Accepted: 4 September 2020; Published: 5 September 2020

Abstract: Many small marine planktonic organisms converge on similar propulsion mechanisms that involve impulsively generated viscous wake vortex rings, and small-scale fluid physics is key to mechanistically understanding the adaptive values of this important behavioral trait. Here, a theoretical fluid mechanics model is developed for plankton jumping, based on observations that the initial acceleration phase for a jumping plankter to attain its maximum speed is nearly impulsive, taking only a small fraction of the viscous timescale, and therefore can be regarded as nearly inviscid, analogous to a one-dimensional elastic collision. Flow circulation time-series data measured by particle image velocimetry (PIV) are input into the model and Froude propulsion efficiencies are calculated for several plankton species. Jumping by the tailed ciliate *Pseudotontonia* sp. has a high Froude propulsion efficiency ~0.9. Copepod jumping also has a very high efficiency, usually >0.95. Jumping by the squid *Doryteuthis pealeii* paralarvae has an efficiency of 0.44 ± 0.16 (SD). Jumping by the small medusa *Sarsia tubulosa* has an efficiency of 0.38 ± 0.26 (SD). Differences in the calculated efficiencies are attributed to the different ways by which these plankters impart momentum on the water during the initial acceleration phase as well as the accompanied different added mass coefficients.

Keywords: plankton jumping; impulsively generated viscous vortex ring; impulsive Stokeslet; impulsive stresslet; elastic collision; Froude propulsion efficiency; added mass coefficient

1. Introduction

Marine planktonic organisms play crucial roles in marine ecosystems and biogeochemical cycling in the world ocean; however, most of them are of microscopic size, having no or limited swimming capabilities relative to the macroscopic water parcels within which they are embedded. Although the water parcels themselves may constantly move in a turbulent, eddying way, the fluid environment at the spatial scales of individual plankters, generally less than a few millimeters, is a dominantly viscous world. Under oceanic turbulence conditions, a copepod is unlikely to face turbulent diffusion inside the spherical space of ~10 mm radius surrounding itself [1], while a small phytoplankton experiences only shear remnants of dissipative turbulent eddies across the space of ~1 mm diameter around itself [2]. Thus, within the fluid immediately surrounding a plankter, the low-Reynolds-number fluid dynamics together with small-scale diffusion governs the transport of mass and momentum, thereby shaping the energy, matter, and information flows to and from the plankter. The small-scale fluid physics is key to mechanistically understanding the adaptations that small marine planktonic organisms engage in to fulfill three main survival tasks, namely feeding, predator avoidance, and reproduction, in the three-dimensional viscous water environment [3–7]. The small-scale fluid physics interfaces with the morphology, behavior, perception, response, and interaction of these small organisms to produce a variety of fascinating phenomena, patterns, processes, and functions that are fundamentally important to marine life, population and ecosystem functioning, and evolution.

Small marine planktonic organisms are morphologically, physiologically, and genetically diverse; however, living in the viscous water environment that is governed by the same small-scale fluid physics has driven convergent evolution of some of their key behavioral traits. For example, the ubiquitous, photosynthetic, jumping ciliate *Mesodinium rubrum* is a species complex that consists of at least six genetically diversified clades [8–10]. Nevertheless, morphologically, they all possess an equatorially located propulsive ciliary belt that enables them to jump both energetically more efficiently and hydrodynamically more quietly [11]; kinematically, they jump at different speeds according to their temperature zones but at the same mean jumping distance of around six body lengths, which is just above the thickness of the nutrient diffusive boundary layer surrounding the cell, indicating the constraint imposed compellingly by the small-scale advection–diffusion physics of the cell's immediately surrounding water [12].

An even more compelling example is that many small marine plankters converge on similar propulsion mechanisms that involve impulsively generated viscous vortex rings. Despite differences in body morphology and size and propulsion machinery, copepods, copepod nauplii, squid paralarvae, small jellyfish, ciliates with contractible tail-like appendages, and freshwater *Daphnia* species generate impulsive viscous vortex rings for fast jumping motions [13–19] or feeding currents [20]. The flow field of an impulsively generated viscous vortex ring can be mathematically described by an impulsive Stokeslet or an impulsive stresslet [13,14], which are spatially limited and temporally ephemeral, thereby effectively reducing the predation risk due to a flow-sensing predator. A jump-imposed flow typically consists of a vortex around the jumping body and a wake vortex, both of which are in a near mirror-image configuration [21], thereby further reducing the predation risk by confusing the real position of the jumping body. Moreover, a relocating jump by a copepod achieves an extremely high Froude propulsion efficiency (>0.9), as revealed by both computational fluid dynamics (CFD) simulations [21] and particle image velocimetry (PIV) measurements [17]. The Froude propulsion efficiency (or hydromechanical efficiency [22]) is defined as

$$\eta \equiv \frac{W_{useful}}{W_{total}} = \frac{\int_0^\tau T(t) U(t) dt}{W_{total}} \qquad (1)$$

where, to reflect the highly unsteady nature of jumping, the useful mechanical work W_{useful} is calculated as the time integral of the product of the instantaneous thrust $T(t)$ and jumping velocity $U(t)$ over the thrust duration τ, and W_{total} is the total mechanical work done for creating the jumping motion. Physically, W_{useful} is the mechanical work that is done to overcome body drag and accelerate the body with added mass to the maximum jumping velocity U_{max}. Although the extremely high Froude propulsion efficiency for copepod jumping appears quite counter-intuitive with respect to the dominantly viscous water environment in which the copepod resides, it simply means that the part of the mechanical work done to generate the wake vortex is significantly smaller than W_{useful}. It is however unknown how the Froude propulsion efficiency varies for jumping by other small plankters that differ in body morphology and size and propulsion machinery.

In the present study, a theoretical fluid mechanics model is developed to calculate the Froude propulsion efficiency for fast jumping by a small plankter. The theoretical model is based on the appropriate assumption that, although the decay phase post the generation of the viscous vortex ring in the wake by an impulsively applied thrust is a highly viscous process, the initial acceleration phase for the body to attain U_{max} is brief and nearly impulsive and therefore can be regarded as a nearly inviscid process. The theoretical model is used to examine the effects on the Froude propulsion efficiency due to such factors as jumping impulsiveness, added mass coefficient, and the different ways by which those small plankters impart momentum on the water during the initial acceleration phase. It is highlighted that, here, the convergent evolution manifests itself in the behavioral traits which those small plankters possess to exert thrust on water in an astonishingly quick and impulsive fashion.

2. The Elastic Collision Model

Here, an elastic collision model is proposed for calculating the Froude propulsion efficiency of impulsive jumping by a small plankter. The theoretical model is developed by dividing the whole jumping process into two phases: an initial acceleration phase, in which the body is self-propelled briefly or impulsively to attain U_{max}, and then a deceleration phase, in which the body decelerates under the action of fluid drag while simultaneously the wake vortex ring initially generated in the acceleration phase decays due to viscosity.

For the acceleration phase, the impulsiveness of jumping ensures that the thrust duration τ is shorter than the viscous timescale defined as $L^2/(4\nu)$, where L is a characteristic length scale (e.g., the body length) and ν is seawater kinematic viscosity, and that the nondimensional jump number defined as

$$N_{jump} = \frac{\tau}{L^2/(4\nu)} \qquad (2)$$

is much smaller than 1. In fact, measurements did show that $N_{jump} \ll 1$ (see below). Formally, τ and L are used as the time and length scale to conduct a dimensional analysis of the vorticity equation $\frac{D\omega}{Dt} = \omega \cdot \nabla u + \nu \nabla^2 \omega$; since $N_{jump} \ll 1$ implies $\frac{\tau}{L^2/\nu} \ll 1$, the viscous diffusion term $\nu \nabla^2 \omega$ is negligible compared with the other two terms. Thus, the acceleration phase is approximately inviscid, and, except for the wake region, the flow around the accelerating body is approximately irrotational. This has two consequences: firstly, the skin drag can be neglected. Secondly, along with accelerating the body of mass m to U_{max}, the body's surface pressure must increase to supply the force to accelerate the fluid around the body, i.e., the added mass m_a must also be accelerated to U_{max} in order to set up the irrotational motion. Therefore, the total impulse of thrust must be $(m + m_a) U_{max}$ [23]. This can be analogous to a one-dimensional elastic collision: the total momentum (=0 at $t = 0$) is conserved in that the maximum momentum achieved by the accelerating body with added mass is equal in magnitude but opposite in direction to the impulse ($\rho_{water} \times I$) of the wake vortex ring that the body generates by impulsively exerting thrust on the water, i.e.,

$$\rho_{water} I = (m + m_a) U_{max} \qquad (3)$$

where ρ_{water} is seawater mass density and I is defined by Equations (A7), (A16), or (A26), respectively, for three types of wake vortex rings that are considered in the following. Moreover, the total mechanical energy that the body expends for propulsion is the addition of the maximum kinetic energy achieved by the accelerating body with added mass [$E \equiv \frac{1}{2}(m + m_a) U_{max}^2$] and the kinetic energy ($\rho_{water} \times K$) of the wake vortex ring, where K is defined by Equations (A8), (A17), or (A27) for the considered three types of wake vortex rings, respectively. Here, the term "elastic collision" is borrowed to mean that the initial separation between the accelerating body and the wake vortex ring conserves both momentum and mechanical energy.

As $m = \rho_{body} V_{body}$ and $m_a = \alpha \rho_{water} V_{body}$ (where ρ_{body} is body mass density, V_{body} is body volume, and α is the added mass coefficient), substituting them into Equation (3) results in an expression for α:

$$\alpha = \frac{I}{V_{body} U_{max}} - \frac{\rho_{body}}{\rho_{water}} \qquad (4)$$

Moreover, from Equation (3), E can be rewritten as

$$E = \frac{\rho_{water}^2 I^2}{2(\rho_{body} + \alpha \rho_{water}) V_{body}} \qquad (5)$$

Then, using E and K, the Froude propulsion efficiency can be calculated as

$$\eta \equiv \frac{E}{E + \rho_{water} K} \qquad (6)$$

For the deceleration phase, the viscous timescale is now the timescale and therefore the viscous diffusion term $\nu \nabla^2 \omega$ is no longer negligible; however, the physical properties of the viscously dissipating wake vortex ring can be used to work out a final formula for calculating the Froude propulsion efficiency. First, a nondimensional kinetic energy of the wake vortex ring is defined as

$$\kappa \equiv \frac{K}{I^{1/2} \Gamma^{3/2}} \tag{7}$$

where Γ is the circulation of the wake vortex ring. Second, substituting Equations (5) and (7) into Equation (6) and using $V_{body} = 4/3 \pi R^3$ (where R is the equivalent spherical radius of the body) lead to

$$\eta = \frac{1}{1 + \frac{8\pi}{3} \kappa \left(\frac{\rho_{body}}{\rho_{water}} + \alpha \right) R^3 I^{-3/2} \Gamma^{3/2}} \tag{8}$$

In Equation (8), I and Γ are informed by PIV measured flow data of the wake vortex ring, and α is calculated using Equation (4) from measured I and U_{max}.

In particular, if the wake vortex ring can be approximated by an impulsive Stokeslet (Appendix A), then it can be shown that

$$\eta_{iStokeslet} = \frac{1}{1 + \frac{\kappa_{iStokeslet}}{3\sqrt{\pi}} \left(\frac{\rho_{body}}{\rho_{water}} + \alpha \right) \left(\frac{R}{\sqrt{\nu t^*}} \right)^3} \tag{9}$$

where $\kappa_{iStokeslet} = \frac{\sqrt{2}}{12} \approx 0.118$ is the nondimensional kinetic energy of the impulsive Stokeslet, and $t^* = t_{max} - t_0$, where t_{max} is the time corresponding to the maximum circulation (Γ_{max}) attained at the end of the acceleration phase (=at the beginning of the deceleration phase), and t_0 is the virtual time origin that is determined by fitting the PIV measured time series of circulation for the deceleration (or decay) phase to the impulsive Stokeslet model (Equation (A6)).

Similarly, if the wake vortex ring can be approximated by one component of the viscous vortex ring pair described by an impulsive stresslet (Appendix B), then it can be shown that

$$\eta_{istresslet} = \frac{1}{1 + \frac{\kappa_{istresslet}}{3\sqrt{\pi}} \left(\frac{\rho_{body}}{\rho_{water}} + \alpha \right) \left(\frac{R}{\sqrt{\nu t^*}} \right)^3} \tag{10}$$

where $\kappa_{istresslet} = \frac{\pi}{20\sqrt{2}} \approx 0.111$ is the nondimensional kinetic energy of the impulsive stresslet, and $t^* = t_{max} - t_0$, where t_{max} is the time corresponding to the maximum circulation (Γ_{max}) and t_0 is the virtual time origin that is determined by fitting the time series of circulation for the decay phase to the impulsive stresslet model (Equation (A15)).

The nondimensional kinetic energy of the wake vortex ring, i.e., κ defined by Equation (7), is a key parameter to describe how energetically costly it is to impose a specific type of wake vortex ring. For a given impulse I and circulation Γ, the higher the value of κ, the higher the mechanical energy cost that is required to generate the specific wake vortex ring and therefore the lower the Froude propulsion efficiency. To illustrate this point, it is hypothetically assumed that a jumping plankter imposes a Hill's spherical vortex (Appendix C) in the wake, and then the resulted Froude propulsion efficiency η_{Hill} is compared with $\eta_{iStokeslet}$ and $\eta_{istresslet}$. It can be shown that

$$\eta_{Hill} = \frac{1}{1 + \frac{10\sqrt{10}\,\kappa_{Hill}}{3\sqrt{\pi}} \left(\frac{\rho_{body}}{\rho_{water}} + \alpha \right) \left(\frac{R}{a} \right)^3} \tag{11}$$

where $\kappa_{Hill} = \frac{\sqrt{10\pi}}{35} \approx 0.160$ is the nondimensional kinetic energy of Hill's spherical vortex, and a is the radius of the vortex. Note that $\kappa_{Hill} > \kappa_{iStokeslet}, \kappa_{istresslet}$.

For real cases of plankton jumping, time-resolved PIV measured circulation data are fitted to the above-mentioned viscous vortex ring models to determine t^*. Results of t^* are then used to calculate the Froude propulsion efficiencies from Equation (9) or (10). These include published data for the copepod *Acartia tonsa* [13,14], the small jellyfish *Sarsia tubulosa* [16], and the tailed ciliate *Pseudotontonia* sp. [18], and previously unpublished data for the squid *Doryteuthis pealeii* paralarvae, the copepod *Calanus finmarchicus*, and the copepod *Acartia hudsonica*. The PIV methods have been adequately described previously [13,20].

3. Results

3.1. General Pattern of Jump-Imposed Flow Fields

Jump-imposed flow fields by different plankton species considered in the present study share a similar general pattern that involves two counter-rotating viscous vortex rings of similar intensity, one in the wake and one around the jumping body (Figure 1; Supplementary Video Group S1). The wake vortex ring is generated by the short duration thrust exerted by the plankter's propulsion machinery on the water. After attaining its maximum speed during a brief accelerating period, the body decelerates due to fluid drag, thereby imparting momentum to the surrounding water and generating the body vortex. Thus, the body vortex lags behind the wake vortex ring roughly by the short duration of thrust.

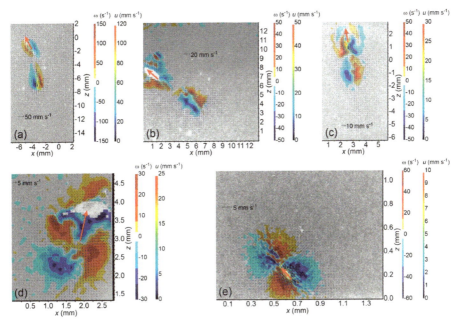

Figure 1. Time-resolved PIV measurements of jump-imposed instantaneous flow velocity vector and vorticity fields by (**a**) a squid *Doryteuthis pealeii* paralarva, (**b**) a copepod *Calanus finmarchicus*, (**c**) a small medusa *Sarsia tubulosa* (PIV data were obtained by [16]), (**d**) a copepod *Acartia hudsonica* female, and (**e**) a ciliate *Pseudotontonia* sp. (PIV data were obtained by [18]). The filled color contours are vorticity fields. The velocity vectors are colored by flow velocity magnitudes. In each panel, the solid red arrow indicates the jumping direction of the body. Details of the events are presented by Supplementary Video Group S1.

3.2. Accleration and Deceleration Phases

For all considered species, their jumping can be divided into an acceleration phase and a deceleration phase (Figure 2). In the acceleration phase, the body accelerates from rest to the maximum speed in a short time that is only a small fraction of the viscous timescale. In the deceleration phase, the body decelerates from its maximum speed under the action of fluid drag; the circulation of the wake vortex ring reaches its maximum but lags slightly behind the body attaining its maximum speed; after this, the wake vortex ring decays immediately due to viscosity.

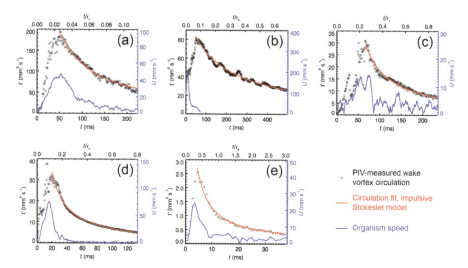

Figure 2. Time evolutions of the time-resolved PIV measured circulations (Γ) of the wake vortex ring, fitted decay-phase circulations (Γ) to the impulsive Stokeslet model, and body speeds (U) for jumps by (**a**) a squid *Doryteuthis pealeii* paralarva, (**b**) a copepod *Calanus finmarchicus*, (**c**) a small medusa *Sarsia tubulosa* (data were obtained by [16]), (**d**) a copepod *Acartia hudsonica* female, and (**e**) a ciliate *Pseudotontonia* sp. (data were obtained by [18]). The cases presented here are respectively corresponding to those presented in Figure 1. Time (*t*) is scaled by the viscous timescale $\tau_v = L^2/(4\nu)$, where L is the body length for (**a**–**d**); for (**e**), L = (body length + tail length)/2. In the initial acceleration phase that is approximately inviscid, the wake vortex circulation increases from 0 at 0 ms to its maximum value roughly at the time τ (i.e., the thrust duration) through the action of thrust and stretching and tilting of vortex lines; thus, the viscous solution of the impulsive Stokeslet model does not apply to this phase. In contrast, in the deceleration phase, the wake vortex circulation stops growing but decays under the action of viscosity; thus, the circulation fit using the impulsive Stokeslet model only covers the decay-phase circulation data starting from the maximum circulation. As to the question of why here the decay-phase circulations are fitted to the impulsive Stokeslet model and not the impulsive stresslet model, previous data analyses have shown that the flow imposed by a relocating copepod *A. tonsa* [14] and by a hopping *Daphnia magna* [19] can be better described by the impulsive stresslet model, while the jumping flow imposed by a small medusa *S. tubulosa* [16] and by a ciliate *Pseudotontonia* sp. [18] can be better described by the impulsive Stokeslet model. Similar data analyses also suggest that the impulsive stresslet model is not a better model for the jumping flow imposed by a squid *D. pealeii* paralarva, by a copepod *C. finmarchicus*, or by a copepod *A. hudsonica* female (not shown for brevity). Thus, here, the decay-phase circulations are fitted to the impulsive Stokeslet model for the five species whose jumping flows are better described by the impulsive Stokeslet model.

3.3. Jump Number

The jumping plankton species considered in this study have a Reynolds number in the range of 3–300. As the Reynolds number decreases, the nondimensional jump number increases but is always kept smaller than 1 (Figure 3). Of particular interest are the ciliates, *Pseudotontonia* sp. and *Tontonia* sp., that have contractible tail-like appendages. They operate at a tail-shrinking time in the range of 2.5–13.4 ms, which is only slightly shorter than the power stroke durations of copepods (6–25 ms). The sizes of the ciliates are at least a few times smaller than those of the copepods; however, the ciliates can increase their effective length scale by deploying a long tail, thereby effectively increasing their viscous timescale to keep their jump number smaller than 1.

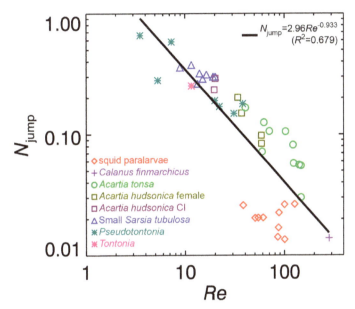

Figure 3. The nondimensional jump number (N_{jump}) as a function of the Reynolds number ($Re \equiv U_{max} L/\nu$, where U_{max} is the maximum jumping speed, L = (body length + tail length)/2 for *Pseudotontonia* and *Tontonia*, and L = body length for all other species). For the copepod *Acartia tonsa*, data were obtained by [13]. For the small medusa *Sarsia tubulosa*, data were obtained by [16]. For the ciliates *Pseudotontonia* sp. and *Tontonia* sp., data were obtained by [18].

3.4. Added Mass Coefficient

The added mass coefficients are calculated from Equation (4) for fast jumping by six plankton species and the calculation results are presented in Table 1. Jumping by the tailed ciliate *Pseudotontonia* sp. has an extremely high added mass coefficient of a mean value of 41.0. Relocating jumping by the copepod *Acartia tonsa* has an added mass coefficient of 0.26 ± 0.24 (SD). An extremely fast jump by a copepod *Calanus finmarchicus* has an added mass coefficient of 1.60 (Example 2 of Supplementary Video Group S1), while an oblique relocating jump by a copepod *A. hudsonica* female has an added mass coefficient of 1.96 (Example 4 of Supplementary Video Group S1). Jumping by the squid *Doryteuthis pealeii* paralarvae has an added mass coefficient of 3.19 ± 1.92 (SD). Jumping by the small medusa *Sarsia tubulosa* has an added mass coefficient of 0.49 ± 0.34 (SD).

Table 1. Summary of the calculation results of the added mass coefficient.

	Pseudotontonia sp.	Acartia tonsa	Acartia hudsonica	Calanus finmarchicus	Squid Paralarvae	Small Sarsia tubulosa
Mean	41.0	0.26	1.96	1.60	3.19	0.49
±SD	-	±0.24	-	-	±1.92	±0.34
Range	37.07–44.92	0.05–0.53	-	-	0.46–6.22	0.12–0.93
n	2	5	1	1	9	4

All calculations assume $\rho_{body}/\rho_{water} = 1$.

3.5. Froude Propulsion Efficiency

The Froude propulsion efficiencies are calculated from Equation (9) or (10) for fast jumping by six plankton species and the calculation results are presented in Table 2. Jumping by the tailed ciliate *Pseudotontonia* sp. has a high Froude propulsion efficiency of 0.904. Copepod jumping also has a high Froude propulsion efficiency, usually >0.95, except for an oblique relocating jump by *Acartia hudsonica* (0.737; Example 4 of Supplementary Video Group S1). Jumping by the squid *Doryteuthis pealeii* paralarvae has a Froude propulsion efficiency of 0.436 ± 0.158 (SD). Jumping by the small medusa *Sarsia tubulosa* has a Froude propulsion efficiency of 0.377 ± 0.259 (SD).

Table 2. Summary of the calculation results of the Froude propulsion efficiency.

	Pseudotontonia sp.	Acartia tonsa [1]	Acartia hudsonica	Calanus finmarchicus [2]	Squid Paralarvae	Small Sarsia tubulosa
Mean	0.904	0.988	0.737	0.953	0.436	0.377
±SD	-	±0.009	-	-	±0.158	±0.259
Range	0.900–0.907	0.974–0.997	-	0.953–0.953	0.187–0.675	0.231–0.764
n	2	5	1	2	9	4

[1] Calculated using Equation (10) for the impulsive stresslet. All other calculations used Equation (9) for the impulsive Stokeslet. [2] An added mass coefficient 1.60 was used for both cases.

Figure 4 shows the functional dependencies of both $\eta_{iStokeslet}$ and $\eta_{istresslet}$ on $\frac{R}{\sqrt{vt^*}}$ and of η_{Hill} on R/a (Equations (9)–(11)). For a given α (e.g., $\alpha = 0.5$; Figure 4), both $\eta_{iStokeslet}$ and $\eta_{istresslet}$ decrease as $\frac{R}{\sqrt{vt^*}}$ increases; since R is the equivalent spherical radius of the body, i.e., a body size scale, while $\sqrt{vt^*}$ is a size scale for the imposed wake vortex ring when it attains its largest circulation, increasing $\frac{R}{\sqrt{vt^*}}$ means decreasing the size of the wake vortex ring at its maximum circulation. Additionally, η_{Hill} decreases as R/a increases; since a is the radius of the Hill's spherical vortex, increasing R/a means decreasing the size of the wake vortex ring.

The Froude propulsion efficiency (η) also depends on the nondimensional kinetic energy of the wake vortex ring (κ). Since $\kappa_{iStokeslet} = \frac{\sqrt{2}}{12} \approx 0.118$ is very close to $\kappa_{istresslet} = \frac{\pi}{20\sqrt{2}} \approx 0.111$, $\eta_{iStokeslet}$ and $\eta_{istresslet}$ are almost identical for given α and $\frac{R}{\sqrt{vt^*}}$ (e.g., $\alpha = 0.5$; Figure 4). Because $\kappa_{Hill} = \frac{\sqrt{10}\pi}{35} \approx 0.160$ is larger than both $\kappa_{iStokeslet}$ and $\kappa_{istresslet}$, η_{Hill} is always smaller than both $\eta_{iStokeslet}$ and $\eta_{istresslet}$ for given α and $\frac{R}{\sqrt{vt^*}} = R/a$ (Figure 4).

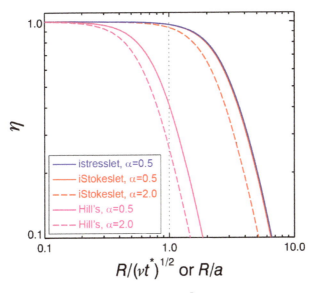

Figure 4. Froude propulsion efficiency η as a function of $\frac{R}{\sqrt{vt^*}}$ based on the impulsive Stokeslet model (Equation (9)) or the impulsive stresslet model (Equation (10)) or as a function of R/a based on the model of Hill's spherical vortex (Equation (11)). For the impulsive Stokeslet model and the model of Hill's spherical vortex, two values of the added mass coefficient are considered, i.e., $\alpha = 0.5$ or 2.0, while for the impulsive stresslet model, $\alpha = 0.5$ is considered. All calculations assume $\rho_{body}/\rho_{water} = 1$.

The Froude propulsion efficiency (η) also depends on the added mass term $\left(\frac{\rho_{body}}{\rho_{water}} + \alpha\right)$ in Equations (8)–(11). For $\frac{\rho_{body}}{\rho_{water}} = 1$, as α increases (e.g., $\alpha = 0.5$ versus $\alpha = 2.0$; Figure 4), $\eta_{iStokeslet}$, $\eta_{istresslet}$, and η_{Hill} all decrease. For $\frac{\rho_{body}}{\rho_{water}} \neq 1$, the denser the body relative to the seawater, the smaller the Froude propulsion efficiency.

4. Discussion

For jumps of a few species considered in this study, the Froude propulsion efficiencies calculated using the newly developed elastic collision model can be compared to available previous results obtained using completely different methods. The Froude propulsion efficiencies calculated in the present study for jumps of the copepods *Acartia tonsa* and *Calanus finmarchicus* (Table 2) compare well to the previous CFD simulation results that range from 0.94 to 0.98 [21]. The present calculation results of Froude propulsion efficiencies for the squid *Doryteuthis pealeii* paralarvae are 0.436 ± 0.158 (SD), ~42% smaller than the previous results of 0.749 ± 0.009 (SD) calculated using PIV-measured jet properties [24]; however, the previous study used 15 Hz for PIV data acquisition, which was not enough to temporally resolve the fast-evolving flow, compared with 1000 Hz for time-resolved PIV in the present study. Nevertheless, these fairly good comparisons suggest that the present elastic collision model captures reasonably well the fluid physics underlying fast jumping motions of many small planktonic organisms. The brief and nearly impulsive acceleration phase for the jumping body to attain its maximum speed can be regarded as a nearly inviscid process, and the jumping body with added mass and its imposed wake vortex ring can be considered as two particles that separate elastically from each other.

The added mass coefficients for jumping by six plankton species have also been estimated from experimental data. To the author's best knowledge, these are the first estimations of added mass coefficients for self-propelled jumping plankters. The estimated added mass coefficients differ generally from that of a towed accelerating sphere (i.e., 0.5), varying significantly for plankton species that differ in size, morphology, and propulsion machinery. The ciliate *Pseudotontonia* sp. deploys a long tail and shrinks it rapidly to accelerate a large amount of the surrounding water, thereby having an extremely high added mass coefficient (~41.0); this extremely high added mass coefficient may be an overestimation but could also be possible because of the big size difference between the cell's main body (~90 μm in length) and the sub-cellular tail that is the propulsion machinery (290–900 μm in length). The relatively high added mass coefficient (~3.19) calculated for the squid *Doryteuthis pealeii* paralarva probably is an overestimation as the effect due to the excess weight of the paralarva should not be neglected in Equation (3). Copepods have a wide range of the added mass coefficient (0.05–1.96), probably because of their teardrop body shapes and versatile jumping behaviors. The highly deformable body shape of the small medusa *Sarsia tubulosa* likely leads to its relatively low added mass coefficient (~0.49). The added mass coefficient is an important parameter that affects the mechanical energy cost and efficiency and imposed flow-field of plankton jumping. The term $\left(\frac{\rho_{body}}{\rho_{water}} + \alpha\right)$ that represents the added-mass effects appears explicitly in Equations (8)–(11) for calculating the Froude propulsion efficiency. The Froude propulsion efficiency decreases as body mass density increases relative to seawater mass density and as the added mass coefficient increases. More research, however, is still needed to shed light on how plankton's morphological and behavioral traits determine their added mass coefficients and in turn how their added mass coefficients affect their individual-level ecological tasks such as jumping and predator–prey interactions.

The Froude propulsion efficiency decreases as the nondimensional kinetic energy of the wake vortex ring increases. A viscous vortex ring (pair) generated by an impulsive Stokeslet (stresslet) has a considerably smaller nondimensional kinetic energy than an inviscid vortex ring such as a Hill's spherical vortex (Equations (7), (A9), (A18), and (A28)). Hence, for given impulse I and circulation Γ, to generate an impulsive viscous vortex ring requires less kinetic energy K than to generate a Hill's spherical vortex. It is not unreasonable to assume that, at the end of the acceleration phase of a jump, the size of the imposed wake vortex ring is similar to the size of the jumping body. For the wake vortex ring to be approximated by an impulsive Stokeslet or stresslet, this assumption means $\frac{R}{\sqrt{vt^*}} = 1$, and therefore the Froude propulsion efficiency $\eta_{iStokeslet} = 0.968$ (Equation (9) and assuming $\alpha = 0.5$) or $\eta_{istresslet} = 0.970$ (Equation (10) and assuming $\alpha = 0.5$). In contrast, for the wake vortex ring to be approximated by a Hill's spherical vortex, this assumption means $R/a = 1$, and therefore $\eta_{Hill} = 0.412$ (Equation (11) and assuming $\alpha = 0.5$), which is significantly lower. Thus, by the very nature of the viscous vortex ring, impulsive jumping by small plankters is an energetically efficient propulsion mode.

The specific way by which a jumping plankter generates a viscous wake vortex ring makes a difference to the Froude propulsion efficiency of its jumping motion. Copepods such as *Acartia tonsa* and *Calanus finmarchicus* beat their swimming legs to generate the viscous wake vortex, and usually they also bend their urosome to aid this process. As a result, their appendage beating movements arrange the imparted momentum on the water in such a way that a well-developed, large viscous vortex ring is generated at the end of the acceleration phase of jumping, thereby achieving $\frac{R}{\sqrt{vt^*}} < 1$ to ensure a high Froude propulsion efficiency. In contrast, being constrained by their evolutionary history, the squid *Doryteuthis pealeii* paralarvae use their jet funnel of ~0.2 mm orifice diameter to deliver momentum on the water. As a result, only a compact, prototype viscous vortex is generated at the end of the acceleration phase, thereby causing $\frac{R}{\sqrt{vt^*}} > 1$ to end up with a relatively low Froude propulsion efficiency. The small jellyfish *Sarsia tubulosa* contracts its bell to initially generate a viscous wake vortex ring; however, likely for feeding purposes, the subsequent bell relaxation draws water into the bell cavity (Example 3 of Supplementary Video Group S1). As a result, the development of the wake vortex ring is arrested to some degree, leading to a low Froude propulsion efficiency. These understandings are visually and qualitatively consistent with available PIV measured instantaneous vorticity fields when circulations reach their maximum.

The overall size of the viscous vortex ring pair generated by the ciliate *Pseudotontonia* sp. is significantly larger than its body size (Figure 1e), because the ciliate boosts its effective length scale by deploying a long, tail-like, sub-cellular structure, and its jumping is accompanied by an extremely high added mass coefficient. This seems to contradict the suggestion that the imposed viscous vortex ring pair helps to reduce the predation risk due to a flow-sensing predator [18]. An alternative explanation may be that the ciliate imposes the viscous vortex ring pair to mimic the presence of copepods that are common predators of protists, thereby scaring away other protistan grazers to reduce feeding competition on algae.

Finally, the present study has applied two Stokes flow models, i.e., the impulsive Stokeslet and stresslet, to plankton jumping flows of Reynolds numbers in the range of 3–300. This is only practically justified by the fact that the decay-phase circulation fits very well to either one of the two models. The authors of [25] investigated partially the effect due to a small, finite Reynolds number on the time evolution of an axisymmetric vortex ring by solving the linearized vorticity equation (i.e., the Stokes approximation). The analytical solution was obtained for an initial condition different from the way a plankton jumping flow is generated, and no closed-form expression is available for the impulse of the vortex ring. Thus, it is difficult to apply this theoretical model to plankton jumping flows. It remains an interesting problem to investigate the finite Reynolds number effects on plankton jumping flows.

Supplementary Materials: The following are available online at http://www.mdpi.com/2311-5521/5/3/154/s1. Supplementary Video Group S1: Five examples of fast jumping motions of plankton that are propelled by generating impulsive viscous vortex rings. Supplementary Video Group S2: Animations of the flow fields and vorticity fields of an impulsive Stokeslet, an impulsive stresslet, and a Hill's spherical vortex.

Funding: This research was funded by US National Science Foundation, grant numbers OCE-1559062 and IOS-1353937, to H.J.

Acknowledgments: The author would like to thank T. Kiørboe, S. Colin, B. Gemmell, E. Buskey, and K. Katija for collaboration on investigating the jumping mechanisms of some of the plankton species that are considered in this manuscript. The author thanks three anonymous reviewers for providing helpful and constructive comments that improved the manuscript.

Conflicts of Interest: The author declares no conflict of interest.

Appendix A. The Impulsive Stokeslet

An impulsive Stokeslet [13] consists of a point momentum source of magnitude ρI, where ρ is the mass density of the fluid and I is the hydrodynamic impulse, acting only impulsively for a very short period of time, formally represented by the Dirac delta function $\delta(t)$. In a cylindrical polar coordinate system (x, r, φ) where x, r, and φ are respectively the axial, radial, and azimuthal coordinate, the azimuthal vorticity (ω_φ) and Stokes streamfunction (Ψ_φ) for the flow imposed by an impulsive Stokeslet are, respectively,

$$\omega_\varphi(x,r,t) = \frac{I\,r}{16\pi^{3/2}\,(\nu t)^{5/2}} e^{-\xi^2} \tag{A1}$$

$$\Psi_\varphi(x,r,t) = \frac{I\,r^2}{2\pi^{3/2}\,(x^2+r^2)^{3/2}} \left(\frac{\sqrt{\pi}}{2} \text{erf}(\xi) - \xi e^{-\xi^2} \right) \tag{A2}$$

where $\xi = \sqrt{\frac{x^2+r^2}{4\nu t}}$, ν is fluid kinematic viscosity, and the error function $\text{erf}(\xi) = \frac{2}{\sqrt{\pi}} \int_0^\xi e^{-y^2} dy$.

The axial (u) and radial (v) velocity components are

$$u(x,r,t) \equiv \frac{1}{r}\frac{\partial \Psi_\varphi}{\partial r} = \frac{I}{2\pi^{3/2}} \frac{(2x^2-r^2)A_1 + 2r^2 A_2}{(x^2+r^2)^{5/2}} \tag{A3a}$$

$$v(x,r,t) \equiv -\frac{1}{r}\frac{\partial \Psi_\varphi}{\partial x} = \frac{I}{2\pi^{3/2}} \frac{xr(3A_1 - 2A_2)}{(x^2+r^2)^{5/2}} \tag{A3b}$$

where $A_1 = \frac{\sqrt{\pi}}{2}\text{erf}(\xi) - \xi e^{-\xi^2}$ and $A_2 = \xi^3 e^{-\xi^2}$. At small time, the far field (i.e., $\xi \gg 1$) of the flow is approximately irrotational (potential) and behaves as

$$u = \frac{I}{4\pi} \frac{2x^2 - r^2}{(x^2 + r^2)^{5/2}} \tag{A4a}$$

$$v = \frac{I}{4\pi} \frac{3xr}{(x^2 + r^2)^{5/2}} \tag{A4b}$$

and the associated velocity magnitude is

$$U_{\text{iStokeslet}} \equiv \sqrt{u^2 + v^2} = \frac{I}{4\pi} \frac{\sqrt{4x^2 + r^2}}{(x^2 + r^2)^2} \tag{A5}$$

The circulation (Γ), impulse (I), and kinetic energy (K) of the viscous vortex ring imposed by the impulsive Stokeslet are, respectively,

$$\Gamma(t) \equiv \int_0^{+\infty} \int_{-\infty}^{+\infty} \omega_\phi dx\, dr = \frac{I}{\pi(4vt)} \tag{A6}$$

$$I(t) \equiv \pi \int_0^{+\infty} \int_{-\infty}^{+\infty} \omega_\phi r^2 dx\, dr = I \tag{A7}$$

$$K(t) \equiv \pi \int_0^{+\infty} \int_{-\infty}^{+\infty} \omega_\phi \Psi_\phi dx\, dr = \frac{\sqrt{2}}{12} \frac{I^2}{\pi^{3/2}(4vt)^{3/2}} \tag{A8}$$

The nondimensional kinetic energy of the viscous vortex ring imposed by the impulsive Stokeslet is

$$K_{\text{iStokeslet}} \equiv \frac{K}{I^{1/2} \Gamma^{3/2}} = \frac{\sqrt{2}}{12} \approx 0.118 \tag{A9}$$

An animation of the flow field and vorticity field of an impulsive Stokeslet is shown in Supplementary Video Group S2.

Appendix B. The Impulsive Stresslet

An impulsive stresslet [14] consists of two point momentum sources of equal magnitude (ρI), acting synchronously in opposite direction and separated by distance ε; each momentum source acts impulsively for a very short period of time, formally represented by $\delta(t)$. The definition of the strength of the impulsive stresslet is $M \equiv \lim_{\varepsilon \to 0, I \to \infty}(I\varepsilon) = $ constant.

The azimuthal vorticity (ω_ϕ) and Stokes streamfunction (Ψ_ϕ) for the flow imposed by an impulsive stresslet are, respectively,

$$\omega_\phi(x, r, t) = \frac{M\, x\, r}{32\pi^{3/2} (vt)^{7/2}} e^{-\xi^2} \tag{A10}$$

$$\Psi_\phi(x, r, t) = \frac{-3M\, x\, r^2}{2\pi^{3/2} (x^2 + r^2)^{5/2}} \left[-\frac{\sqrt{\pi}}{2} \text{erf}(\xi) + \xi e^{-\xi^2}\left(1 + \frac{2}{3}\xi^2\right) \right] \tag{A11}$$

The axial (u) and radial (v) velocity components are

$$u(x, r, t) \equiv \frac{1}{r} \frac{\partial \Psi_\phi}{\partial r} = \frac{M}{2\pi^{3/2}} \frac{x[(2x^2 - 3r^2)B_1 + r^2 B_2]}{(x^2 + r^2)^{7/2}} \tag{A12a}$$

$$v(x, r, t) \equiv -\frac{1}{r} \frac{\partial \Psi_\phi}{\partial x} = -\frac{M}{2\pi^{3/2}} \frac{r[(r^2 - 4x^2)B_1 + x^2 B_2]}{(x^2 + r^2)^{7/2}} \tag{A12b}$$

where $B_1 = 3\left[-\frac{\sqrt{\pi}}{2}\text{erf}(\xi) + \xi e^{-\xi^2}\left(1 + \frac{2}{3}\xi^2\right)\right]$ and $B_2 = -4\xi^5 e^{-\xi^2}$. At small time, the far field (i.e., $\xi \gg 1$) of the flow is approximately irrotational (potential) and behaves as

$$u = \frac{3M}{4\pi} \frac{x(2x^2 - 3r^2)}{(x^2 + r^2)^{7/2}} \tag{A13a}$$

$$v = \frac{3M}{4\pi} \frac{r(4x^2 - r^2)}{(x^2 + r^2)^{7/2}} \tag{A13b}$$

and the associated velocity magnitude is

$$U_{\text{istresslet}} \equiv \sqrt{u^2 + v^2} = \frac{3M}{4\pi} \frac{\sqrt{4x^4 + r^4}}{(x^2 + r^2)^3} \tag{A14}$$

The circulation (Γ), impulse (I), and kinetic energy (K) of each component of the viscous vortex ring pair imposed by the impulsive stresslet are, respectively,

$$\Gamma(t) \equiv \int_0^{+\infty}\int_0^{+\infty} \omega_\phi dx\, dr = \frac{M}{\pi^{3/2}(4vt)^{3/2}} \tag{A15}$$

$$I(t) \equiv \pi \int_0^{+\infty}\int_0^{+\infty} \omega_\phi r^2 dx\, dr = \frac{M}{\pi^{1/2}(4vt)^{1/2}} \tag{A16}$$

$$K(t) \equiv \pi \int_0^{+\infty}\int_0^{+\infty} \omega_\phi \Psi_\phi dx\, dr = \frac{1}{20\sqrt{2}} \frac{M^2}{\pi^{3/2}(4vt)^{5/2}} \tag{A17}$$

The nondimensional kinetic energy of each component of the viscous vortex ring pair imposed by the impulsive stresslet is

$$\kappa_{\text{istresslet}} \equiv \frac{K}{I^{1/2}\Gamma^{3/2}} = \frac{\pi}{20\sqrt{2}} \approx 0.111 \tag{A18}$$

An animation of the flow field and vorticity field of an impulsive stresslet is shown in Supplementary Video Group S2.

Appendix C. Hill's Spherical Vortex

Hill's spherical vortex is an axisymmetric vortex ring with the strength of the vorticity within the vortex proportional to the distance from the axis of traveling (e.g., [26,27]). That is, in a cylindrical polar coordinate system (x, r, φ) with the positive axial (x-) direction coinciding with the traveling direction, the azimuthal vorticity is given by $\omega_\varphi = C r$, where C is a constant. Outside the vortex, the vorticity vanishes, and the flow is identical to that for uniform irrotational flow due to the translation of a sphere.

In a stationary frame of reference, the Stokes streamfunction inside and outside Hill's spherical vortex of radius a is given by

$$\Psi_{\text{in}}(x, r, t) = \frac{U}{2}r^2 + \frac{C}{10}r^2\left[a^2 - (x - Ut)^2 - r^2\right] \tag{A19}$$

$$\Psi_{\text{out}}(x, r, t) = \frac{U}{2}r^2 - \frac{U}{2}r^2\left\{1 - \frac{a^3}{\left[(x - Ut)^2 + r^2\right]^{3/2}}\right\} \tag{A20}$$

where $U = 2Ca^2/15$, which is the vortex translation velocity.

Within the vortex, the axial (u_{in}) and radial (v_{in}) velocity components are

$$u_{in}(x,r,t) \equiv \frac{1}{r}\frac{\partial \Psi_{in}}{\partial r} = U + \frac{C}{5}\left[a^2 - (x-Ut)^2 - 2r^2\right] \tag{A21}$$

$$v_{in}(x,r,t) \equiv -\frac{1}{r}\frac{\partial \Psi_{in}}{\partial x} = \frac{C}{5}r(x-Ut) \tag{A22}$$

Outside the vortex, the axial (u_{out}) and radial (v_{out}) velocity components are

$$u_{out}(x,r,t) \equiv \frac{1}{r}\frac{\partial \Psi_{out}}{\partial r} = Ua^3 \frac{2(x-Ut)^2 - r^2}{2\left[(x-Ut)^2 + r^2\right]^{5/2}} \tag{A23a}$$

$$v_{out}(x,r,t) \equiv -\frac{1}{r}\frac{\partial \Psi_{out}}{\partial x} = Ua^3 \frac{3r(x-Ut)}{2\left[(x-Ut)^2 + r^2\right]^{5/2}} \tag{A23b}$$

and the associated velocity magnitude is

$$U_{out} \equiv \sqrt{u_{out}^2 + v_{out}^2} = Ua^3 \frac{\sqrt{4(x-Ut)^2 + r^2}}{2\left[(x-Ut)^2 + r^2\right]^2} \tag{A24}$$

The circulation (Γ), impulse (I), and kinetic energy (K) of Hill's spherical vortex are, respectively,

$$\Gamma \equiv \int_0^{+\infty} \int_{-\infty}^{+\infty} \omega_\phi dx\, dr = 5Ua \tag{A25}$$

$$I \equiv \pi \int_0^{+\infty} \int_{-\infty}^{+\infty} \omega_\phi r^2 dx\, dr = 2\pi a^3 U \tag{A26}$$

$$K \equiv \pi \int_0^{+\infty} \int_{-\infty}^{+\infty} \omega_\phi \Psi dx\, dr = \frac{10}{7}\pi a^3 U^2 \tag{A27}$$

The nondimensional kinetic energy of Hill's spherical vortex is

$$\kappa_{Hill} \equiv \frac{K}{I^{1/2}\Gamma^{3/2}} = \frac{\sqrt{10\pi}}{35} \approx 0.160 \tag{A28}$$

An animation of the flow field and vorticity field of a Hill's spherical vortex is shown in Supplementary Video Group S2.

References

1. Osborn, T. The role of turbulent diffusion for copepods with feeding currents. *J. Plankton Res.* **1996**, *18*, 185–195. [CrossRef]
2. Lazier, J.R.N.; Mann, K.H. Turbulence and diffusive layers around small organisms. *Deep Sea Res.* **1989**, *36*, 1721–1733. [CrossRef]
3. Koehl, M.A.R.; Strickler, J.R. Copepod feeding currents: Food capture at low Reynolds number. *Limnol. Oceanogr.* **1981**, *26*, 1062–1073. [CrossRef]
4. Strickler, J.R. Sticky water: A selective force in copepod evolution. In *Trophic Interactions within Aquatic Ecosystems*; Meyers, D.G., Strickler, J.R., Eds.; AAAS Selected Symposium 85, American Association for the Advancement of Science; Westview Press: Boulder, CO, USA, 1984; pp. 187–239.
5. Yen, J.; Strickler, J.R. Advertisement and concealment in the plankton: What makes a copepod hydrodynamically conspicuous? *Invertebr. Biol.* **1996**, *115*, 191–205. [CrossRef]
6. Jiang, H.; Osborn, T.R. Hydrodynamics of copepods: A review. *Surv. Geophys.* **2004**, *25*, 339–370. [CrossRef]
7. Kiørboe, T. *A Mechanistic Approach to Plankton Ecology*; Princeton University Press: Princeton, NJ, USA, 2008.

8. Herfort, L.; Peterson, T.D.; McCue, L.A.; Crump, B.C.; Prahl, F.G.; Baptista, A.M.; Campbell, V.; Warnick, R.; Selby, M.; Roegner, G.C.; et al. *Myrionecta rubra* population genetic diversity and its cryptophyte chloroplast specificity in recurrent red tides in the Columbia River estuary. *Aquat. Microb. Ecol.* **2011**, *62*, 85–97. [CrossRef]
9. Garcia-Cuetos, L.; Moestrup, Ø.; Hansen, P.J. Studies on the genus *Mesodinium* II. Ultrastructural and molecular investigations of five marine species help clarifying the taxonomy. *J. Eukaryot. Microbiol.* **2012**, *59*, 374–400. [CrossRef] [PubMed]
10. Johnson, M.D.; Beaudoin, D.J.; Laza-Martinez, A.; Dyhrman, S.T.; Fensin, E.; Lin, S.; Merculief, A.; Nagai, S.; Pompeu, M.; Setälä, O.; et al. The genetic diversity of *Mesodinium* and associated cryptophytes. *Front. Microbiol.* **2016**, *7*, 2017. [CrossRef]
11. Jiang, H. Why does the jumping ciliate *Mesodinium rubrum* possess an equatorially located propulsive ciliary belt? *J. Plankton Res.* **2011**, *33*, 998–1011. [CrossRef]
12. Jiang, H.; Johnson, M.D. Jumping and overcoming diffusion limitation of nutrient uptake in the photosynthetic ciliate *Mesodinium rubrum*. *Limnol. Oceanogr.* **2017**, *62*, 421–436. [CrossRef]
13. Kiørboe, T.; Jiang, H.; Colin, S.P. Danger of zooplankton feeding: The fluid signal generated by ambush-feeding copepods. *Proc. R. Soc. B* **2010**, *277*, 3229–3237. [CrossRef] [PubMed]
14. Jiang, H.; Kiørboe, T. The fluid dynamics of swimming by jumping in copepods. *J. R. Soc. Interface* **2011**, *8*, 1090–1103. [CrossRef] [PubMed]
15. Murphy, D.W.; Webster, D.R.; Yen, J. A high-speed tomographic PIV system for measuring zooplanktonic flow. *Limnol. Oceanogr. Methods* **2012**, *10*, 1096–1112. [CrossRef]
16. Katija, K.; Jiang, H. Swimming by medusae *Sarsia tubulosa* in the viscous vortex ring limit. *Limnol. Oceanogr. Fluids Environ.* **2013**, *3*, 103–118. [CrossRef]
17. Wadhwa, N.; Andersen, A.; Kiørboe, T. Hydrodynamics and energetics of jumping copepod nauplii and copepodids. *J. Exp. Biol.* **2014**, *217*, 3085–3094. [CrossRef]
18. Gemmell, B.J.; Jiang, H.; Buskey, E.J. A tale of the ciliate tail: Investigation into the adaptive significance of this sub-cellular structure. *Proc. R. Soc. B* **2015**, *282*, 20150770. [CrossRef]
19. Skipper, A.N.; Murphy, D.W.; Webster, D.R. Characterization of hop-and-sink daphniid locomotion. *J. Plankton Res.* **2019**, *41*, 142–153. [CrossRef]
20. Jiang, H.; Paffenhöfer, G.-A. Vortical feeding currents in nauplii of the calanoid copepod *Eucalanus pileatus*. *Mar. Ecol. Prog. Ser.* **2020**, *638*, 51–63. [CrossRef]
21. Jiang, H.; Kiørboe, T. Propulsion efficiency and imposed flow fields of a copepod jump. *J. Exp. Biol.* **2011**, *214*, 476–486. [CrossRef]
22. Lighthill, M.J. Aquatic animal propulsion of high hydromechanical efficiency. *J. Fluid Mech.* **1970**, *44*, 265–301. [CrossRef]
23. Lighthill, J. *An Informal Introduction to Theoretical Fluid Mechanics*; Clarendon Press: Oxford, UK, 1986.
24. Bartol, I.K.; Krueger, P.S.; Stewart, W.J.; Thompson, J.T. Pulsed jet dynamics of squid hatchlings at intermediate Reynolds numbers. *J. Exp. Biol.* **2009**, *212*, 1506–1518. [CrossRef] [PubMed]
25. Fukumoto, Y.; Kaplanski, F. Global time evolution of an axisymmetric vortex ring at low Reynolds numbers. *Phys. Fluids* **2008**, *20*, 053103. [CrossRef]
26. Lim, T.T.; Nickels, T.B. Vortex rings. In *Fluid Vortices*; Green, S.I., Ed.; Kluwer Academic Publishers: Dordrecht, The Netherlands, 1995; pp. 95–153.
27. Pozrikidis, C. *Introduction to Theoretical and Computational Fluid Dynamics*; Oxford University Press: Oxford, UK, 1997.

 © 2020 by the author. Licensee MDPI, Basel, Switzerland. This article is an open access article distributed under the terms and conditions of the Creative Commons Attribution (CC BY) license (http://creativecommons.org/licenses/by/4.0/).

Review

Kinematic and Dynamic Scaling of Copepod Swimming

Leonid Svetlichny [1,*], Poul S. Larsen [2] and Thomas Kiørboe [3]

[1] I.I. Schmalhausen Institute of Zoology, National Academy of Sciences of Ukraine, Str. B. Khmelnytskogo, 15, 01030 Kyiv, Ukraine
[2] DTU Mechanical Engineering, Fluid Mechanics, Technical University of Denmark, Building 403, DK-2800 Kgs. Lyngby, Denmark; psl@mek.dtu.dk
[3] Centre for Ocean Life, Danish Technical University, DTU Aqua, Building 202, DK-2800 Kgs. Lyngby, Denmark; tk@aqua.dtu.dk
* Correspondence: leonid.svetlichny@gmail.com

Received: 30 March 2020; Accepted: 6 May 2020; Published: 11 May 2020

Abstract: Calanoid copepods have two swimming gaits, namely cruise swimming that is propelled by the beating of the cephalic feeding appendages and short-lasting jumps that are propelled by the power strokes of the four or five pairs of thoracal swimming legs. The latter may be 100 times faster than the former, and the required forces and power production are consequently much larger. Here, we estimated the magnitude and size scaling of swimming speed, leg beat frequency, forces, power requirements, and energetics of these two propulsion modes. We used data from the literature together with new data to estimate forces by two different approaches in 37 species of calanoid copepods: the direct measurement of forces produced by copepods attached to a tensiometer and the indirect estimation of forces from swimming speed or acceleration in combination with experimentally estimated drag coefficients. Depending on the approach, we found that the propulsive forces, both for cruise swimming and escape jumps, scaled with prosome length (L) to a power between 2 and 3. We further found that power requirements scales for both type of swimming as L^3. Finally, we found that the cost of transportation (i.e., calories per unit body mass and distance transported) was higher for swimming-by-jumping than for cruise swimming by a factor of 7 for large copepods but only a factor of 3 for small ones. This may explain why only small cyclopoid copepods can afford this hydrodynamically stealthy transportation mode as their routine, while large copepods are cruise swimmers.

Keywords: copepods; cruising; escape swimming; kinematics; hydrodynamics; power; cost of transport

1. Introduction

The swimming of pelagic copepods is based on the principle of rowing strokes with oar-like limbs. The anatomy of the body structure is directly related to the way of swimming, and copepods are divided into two main groups: the ancient Gymnoplea and the more recent Podoplea [1,2]. In Gymnoplea, which includes the Calanoida, both the cephalic and thoracic limbs participate in propulsion. The cephalic appendages perform the combined functions of feeding and steady cruise swimming [3]. In Podoplea, only the thoracic limbs—the swimming legs—are involved in swimming. The thoracic limbs in all copepods, with the exception of some parasitic taxa, are used for jumping.

The first descriptions of the kinematics of the cephalic appendages of copepods belonged to Storch and Pfisterer [4] and Cannon [3]. Subsequently, they were supplemented by Lowndes [5] and developed by Gauld [6] and Petipa [7]. The purpose of these experimental works was to elucidate the copepod feeding mechanisms, and they were performed using filming, polygraphs, and stroboscopic photography. The concept of filtration feeding was developed based on these studies. More advanced high-speed filming later revealed that the feeding mechanism is not the filtering of particles through a sieve; rather, the feeding current is a scanning current [8]. The use of high-speed filming made it possible to reveal new details of the complex interaction of the cephalic limbs during feeding and movement, and it was demonstrated that the frequency of cephalic limb beating in copepods varies between 20 and 40 Hz but can reach 70–80 Hz [9–12]. Thus, even during slow swimming, the limbs oscillate so fast that analyzing their action requires video recordings with a frequency of 700–800 Hz to obtain a good resolution of the leg stroke phase. During escape swimming, the requirements for recording frequency are even higher because limb frequencies may be as high as 200 Hz [13].

Storch [14] may have been the first to use a high-speed movie camera at 120 frames per second to study the jumping behavior of freshwater cyclopoid copepods. He described the metachronal strokes of the thoracic legs of *Cyclops scutifer* during avoidance response. Subsequent studies, using increasingly higher frame rates of up >3000 fps, estimated incredibly high swimming speeds during escape jumps of >500 body lengths per second, and they provided detailed descriptions of the movement of the feeding appendages and swimming legs during cruise swimming and jumps [7,15–22]. These high resolution observations of swimming speeds and appendage kinematics provided the basis for estimations of the force production and energetics of copepod propulsion [23–28]. From observations of speed or acceleration, together with estimates of drag of the moving body or limbs, it is possible to estimate force production.

An alternative approach to estimate force production during swimming and jumping is to directly measure forces of animals tethered to a tensiometer [29–32], a spring [33], or an aluminum wire whose deflection is calibrated and monitored by a displacement sensor [34,35].

The aim of this synthesis was to describe limb kinematics and examine the magnitudes and size scaling of force production and energy expenditure during cruise and jump swimming in copepods. We combined available literature data with our own new data on swimming speed, appendage kinematics, drag measurements, force measurements on attached specimens, and direct and indirect estimates of force production. We analyzed observations by means of simple theoretical models, and we provide correlations that reflect size scaling laws for kinematics, force, power, and drag. All symbols used are listed in Table 1.

Table 1. List of symbols.

a	acceleration
c	hydrodynamic shape factor
C_d	coefficient of drag
C_t	energy consumption per unit body mass and time
D	diameter of body
D	duration
E	energy
F	frequency of beat
K	empirical constant
l_a	effective length of second antenna
L	prosome length
M	body mass
N	power, energy per unit time
Re	Reynols tal, $\rho L U/\mu$
R_d	drag force
R_p	propulsive force
S	sectional area of body
S	distance
S_{loc}	locomotor step length
U	body speed
U_a	circular speed of second antenna

Greek	
α	angle of second antenna beat
μ	dynamic viscosity
ν	kinematic viscosity, μ/ρ
ρ	density
Subscripts	
att	attached, tethered to force sensor
cr	cruising, free
d	drag
esc	escape jump
kick	kick, jump
max	maximal
mean	mean
min	minimal
p	propulsion
st	stroke phase

Table 1. *Cont.*

2. Locomotor Function of Appendages

2.1. Cruise Swimming

The cephalic appendages serve the functions of propulsion and the capture of food particles. Depending on the degree to which the cephalic appendages combine these functions, one can identify three main kinematics (Figure 1). For an older group of cruising feeders (Figure 1A), such as *Calanus*, *Paracalanus* and *Pseudocalanus* that consume food particles during continuous uniform swimming, the main feature of their limb movement is the antiphase action of the second antennas and maxillipeds [3,5,10].

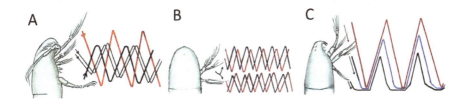

Figure 1. Schematic representation of the action of cephalic appendages in calanoid copepods in terms of their angular movements during cruise swimming. Each line starts from the nearest drawn cephalic appendages. (**A**): *Calanus helgolandicus*. The red and two thin black lines directed downward correspond to the angular movement of second antennas, mandibles, and maxillas; the blue line directed up corresponds to the movement of maxillipeds (from [12]). (**B**): *Eurytemora affinis*. The upper red and blue lines show the movement of exopodite and endopodite, respectively, of the second antenna; the lower red and blue lines correspond to the movement of the exopodite and endopodite, respectively, of the mandible. (**C**): *Anomalocera patersoni*. Red, blue, and black lines correspond to movement of second antennas, mandibles, and maxillas, respectively.

The limb kinematics determines the resulting propulsive force, which allows the copepods to swim steadily (Figure 2A). This is evidenced by experiments with the amputation of individual pairs of cephalic appendages. After the amputation of the maxillipeds, the force resulting from the partially antiphase action of the second antennas, mandibles, and maxillas has been found to remain the same, but a pronounced inverse component of the force has been found to appear (Figure 2B). It was found that the amplitude of the force of the second antennae alone is higher again than the force resulting individuals, since all copepods, except for some phases of their ontogenesis, are negatively buoyant [45]. Some studies have suggested that in *Eurytemora affinis*, the antennae are active contributors to the production of propulsive force [46]. However, numerous high-speed studies of relocation jumps have shown that the antennules are pressed close to the body during the first power stroke of a jump event, and they then remain passive during subsequent power strokes [16,19,30,47]. An exception is the swimming of males of the genus *Oithona* that swim due to power strokes of almost all limbs, including short antennae, cephalic appendages, thoracic limbs, and the abdomen [47,48].

3. Scaling of Swimming Kinematics

3.1. Cruising of Calanoid Copepods

To identify the large-scale patterns of cruise swimming copepods, we used our own and published data obtained by high-speed methods to simultaneously determine the swimming speed (U, cm s^{-1}) and beat frequency of cephalic appendages (F, Hz) as a function of the body length (L, cm) of individual specimens (Table 2). Swimming speed increases with body length to a power of approximately 1.4; 'the locomotor step length' (S_{loc}), i.e., the distance that the copepod covers during one beat cycle, increases approximately with the square of the body length; and limb beat frequency varies approximately inversely with the square root of body length (Figure 4).

Table 2. Kinematic parameters of cruise swimming calanoid copepods at 20 °C. L_{pr}: prosome length; L_{an}: effective length of second antenna, measured as the distance from body to mid-area of marginal bristles of endopodites; n_{ind}: number of individuals; n_m: number of measurements; F: frequency of cephalic appendages at cruising speed; U: horizontal body speed; $S_{loc} = U_{body}/F$: locomotor step.

Species	L (cm)	n_{ind}/n_m	F (Hz)	U (cm s^{-1})	S_{loc} (cm)	Source
Paracalanus parvus	0.063	8/56	63.9 ± 12.4	0.31 ± 0.15	0.005	
	-"-	3/8	75.9 ± 5.3	0.8 ± 0.25	0.011	
Acartia tonsa	0.084	8/86	77.8 ± 4.6	0.33 ± 0.4	0.004	
	-"-	4/32	66.0 ± 5.1	0.4 ± 0.6	0.006	
Centropages ponticus	0.084	6/34	69.0 ± 8	0.45 ± 0.13	0.007	
Pseudocalanus elongatus	0.084	7/24	41.8 ± 7.3	0.56 ± 0.27	0.013	
Euritemora affinis	0.08	2/54	68.4 ± 3.2	0.64 ± 0.29	0.009	
	-"-	1/3	66.7 ± 2.4	0.45 ± 0.13	0.007	
Centropages typicus	0.112	11/109	39.6 ± 4.1	0.81 ± 0.38	0.020	
	-"-	1/4	42.7 ± 1.3	1.37 ± 0.33	0.032	Present data
Limnocalanus macrurus	0.18	4/40	41.7 ± 5.5	0.84 ± 0.09	0.020	
	-"-	1/5	39.7 ± 4.3	0.53 ± 0.01	0.013	
Pontella mediterranea	0.20	5/26	23.4 ± 1.2	3.1 ± 0.57	0.132	
	-"-	2/8	26.3 ± 1.7	2.5 ± 0.8	0.153	
Calanus helgolandicus	0.27	7/82	36.0 ± 2.7	2.16 ± 0.45	0.060	
	-"-	4/9	41.3 ± 5.2			
Anomalocera patersoni	0.25	3/38	26.4 ± 10.1	3.5 ± 1.7	0.133	
	-"-	6/26	21.3 ± 4.1			

Table 2. Cont.

Species	L (cm)	n_{ind}/n_m	F (Hz)	U (cm s^{-1})	S_{loc} (cm)	Source
Pseudodiaptomus marinus	0.082	5/39	80.4 ± 6.8	0.24 ± 0.06	0.003	
Paracalanus parvus	0.06		63.0 ± 6	0.35 ± 0.05	0.006	
	0.063		72.3 ± 4			
Pseudocalanus elongatus	0.08		45.2 ± 5	0.48 ± 0.17	0.011	
Centropages ponticus	0.086		64.0 ± 1			
Acartia clausi	0.095		51.7 ± 10			
Pontella mediterranea	0.24		27 ± 3			[32]
Calanus helgolandicus	0.26		39.1 ± 5	2.69 ± 0.1	0.068	
Neocalanus gracilis	0.25		28.0 ± 2			
Euchirella messinensis	0.35		29.1			
Euchaeta marina	0.3		55.0 ± 5			
Pleuromamma abdominalis	0.23		37 ± 3			
Phaenna spinifera	0.14		59.5 ± 3			
Calanus helgolandicus	0.27			3.2		
Rhincalanus nasutus	0.5			0.59		
Euchirella curticauda	0.36			2		[49]
Euchaeta marina	0.33			2.5		
Scolecthrix sp,	0.18			1.1		
Anomalocera patersoni	0.31			5.32		
Diaptomus kenai	0.18			0.5 ± 0.1		
Diaptomus tyrelli	0.08			0.05		[50]
Diaptomus hesperus	0.15		50	0.31	0.006	
Eucalanus pileatus	0.14		18			[51]
Paracalanus parvus	0.07		83			
Centropages typicus	0.14		55			[52]
Calanus sinicus	0.23			1.14		[53]
Temora longocornis	0.09		32. ± 3			[54]
Eurytemora hirundoides	0.084			0.34		[55]
Acartia granii (females)	0.101			0.33 ± 0.5		
Temora longicornis (females)	0.074			0.14 ± 0.19		
Temora stylifera (females)	0.107			0.33 ± 0.35		
Pseudocalanus elongatus (females)	0.079			0.2 ± 0.26		[48]
Acartia granii (males)	0.088			0.34 ± 0.84		
Temora longicornis (males)	0.068			0.3 ± 0.23		
Temora stylifera (males)	0.099			0.72 ± 0.46		
Pseudocalanus elongatus (males)	0.064			0.28 ± 0.3		
Temora longicornis	0.085		40.7 ± 8	0.48 ± 0.9		[17]
Centropages velificatus	0.12			0.7		[56]
Paracalanus aculeatus	0.1			0.2		
Euchaeta rimana	0.25			0.75 ± 0.04		[57]

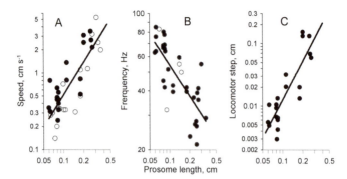

Figure 4. (**A**): Regressions of average speed (U). (**B**): Limb beat frequency (F). (**C**): 'Locomotor step length' (S_{loc}) versus prosome length (L) during cruise swimming (data from Table 2). Black circles are own data obtained from 1200 fps videos. Empty circles are literature data. The power-law regressions were based on all data, $U = 13.4\, L^{1.4}$ ($R^2 = 0.69$); $F = 16.0\, L^{-0.53}$ ($R^2 = 0.59$); $S_{loc} = 1.36\, L^{2.03}$ ($R^2 = 0.73$).

3.2. Kinematic Analysis of Escape Reaction

Since even modern high-speed cameras do not allow for long-term recordings of animal activity, the copepod escape reaction may be synchronized with video records by various external means of stimulation, such as short, weak electrical pulses [30,31] or photic and hydrodynamic stimuli [19,20,58]. In our studies, we used short electrical impulses (see [30,31]). With this dosage, we observed a stable and maximum motor response. Another advantage is that all the copepod species studied by us showed positive galvanotaxis. With the lateral placement of the electrodes, this increased the likelihood of individuals moving in the focal plane of the camera lens, therefore providing sharper images. After each period of stimulation, the copepods were replaced with new animals.

Video sequences showing specimens moving in the focal plane were selected for frame-by-frame analysis. We digitized the geometric center of the prosome of the copepod and computed velocities from the change of this position between frames. Video recording was performed at 1200 fps with a back collimated beam of light from a 5 W LED lamp. All measured parameters describing the kinematics of the escape reaction are explained in the Supplementary Table S1.

It has been previously shown that the direction of trajectory can change dramatically, even up to a complete turn, during a power stroke [19,28].

However, even during rectilinear movement, power strokes by the abdomen and swimming legs cause a dorsal rotation of the body, while returning the limbs to their original position leads to the rotation of the body in the ventral direction [31]. Particularly pronounced are such body rotations in copepods with elongated abdomens. For example, in the cyclopoid copepod *Oithona davisae* with a total body length of 0.05 cm, the ventral deviation of the body axis from the direction of movement at the end of the kick has been found to reach 90°. From this position, the next kick starts (Figure 5).

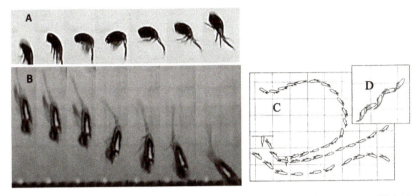

Figure 5. Instantaneous body positions of *Oithona davisae* (**A**) and *Limnocalanus macrurus* (**B**) during escape kick, the trajectory of three individuals of *Calanus helgolandicus* stimulated by electrical impulses (**C**), and the instantaneous positions of the body at the end of stroke and recovery phases of kick (**D**). Note, only characteristic body positions are shown in (**A**,**B**) (present data). (**C**,**D**) are from [31].

In larger (0.3 cm) calanoid copepods of *Limnocalanus macrurus* with a very long abdomen, the turning of the body axis has been found to reach 45°. For copepods with a relatively short abdomen, such as *Paracalanus* and *Calanus*, the angular amplitude of oscillations of the body axis relative to the direction of motion is about 30°. Nevertheless, in small *Acartia tonsa* (<0.1 cm), the body angle can vary within 55° [13]. All copepods also rotate their body around their longitudinal axes [19,31].

3.2.1. Instantaneous and Average Speed of Escape Reaction

A complete escape reaction is made up of a series of kicks [22,31,59,60]. During the inertial phase between kicks, the velocity decreases to U_{min} immediately before the next kick. In the smallest *Oithona davisae* and *Oithona nana* (~0.03 cm prosome length), the average U_{min} was 2.8 ± 1.4 cm s^{-1} (Figure 6) and increased in large (0.28–0.39 cm) species to 28.7 ± 9.7 in *Calanus helgolandicus*, 45.0 ± 15.6 cm s^{-1} in *Euchirella messinensis* [60], and about 40 cm s^{-1} in *Calanus finmarchicus* [22].

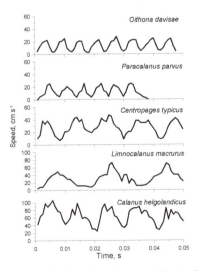

Figure 6. Instantaneous speeds of 5 species of copepods during the escape reaction.

It has previously been shown that both the maximum and average speed of escape reaction correlate with the size of the copepod body [20–22,28,60–63]. Our new data included the results of the video recording (1200 fps) of the escape reaction of 15 species of copepods and updated results of the old filming (3000 fps) of the escape reaction of the larger Mediterranean copepods *Euchaeta media* and *Euchirella messinensis* (Table 3).

Table 3. Kinematic parameters of the escape reaction in calanoid and cyclopoid copepods at 20–22 °C. L_{pr}: prosome length, cm; U_{max}: maximum instantaneous speed, cm s^{-1}; U_{kick}: mean speed of kick, cm s^{-1}; D_{kick}: total duration of kick, s; S_{kick}: total distance of kick, cm; N: number of measurements. Average values are means ± standard deviation. The literature data included in the table were obtained with a high-speed registration of at least 500 fps.

Species	L_{pr}, cm	N	U_{max}, cm s^{-1}	U_{kick}, cm s^{-1}	D_{kick}, s	S_{kick}, cm	Source
Oithona davisae	0.028	41	17.5 ± 6.3	10.0 ± 3.7	0.0081 ± 0.0023	0.065 ± 0.016	
Oithona nana	0.031	25	21.4 ± 2.5	10.1 ± 1.2	0.0076 ± 0.0009	0.074 ± 0.012	
Oithona similis	0.045	35		12.1 ± 2.3	0.0077 ± 0.0011	0.093 ± 0.014	
Paracalanus parvus	0.06	30	20.8 ± 3.9	11.9 ± 2.5	0.0066 ± 0.0011	0.077 ± 0.013	
Pseudodiaptomus marinus	0.082	17	56.6 ± 7.7	31.9 ± 3.9	0.0075 ± 0.0008	0.238 ± 0.033	
Eurytemora affinis	0.08	13	38.7 ± 5.2	21.9 ± 2.7	0.0083 ± 0.0012	0.182 ± 0.028	
Acartia clausi	0.089	29	48.3 ± 9.9	28.1 ± 6.0	0.0062 ± 0.0013	0.170 ± 0.039	
Acartia tonsa	0.085	9	54.5 ± 4.4	30.2 ± 3.2	0.0059 ± 0.0008	0.176 ± 0.022	Present data
Centropages ponticus	0.084	5	27.2 ± 8.1	16.9 ± 4.7	0.0105 ± 0.0004	0.177 ± 0.052	
Pseudocalanus elongatus	0.086	17	36.0 ± 4.5	19.8 ± 2.8	0.0082 ± 0.0010	0.163 ± 0.037	
Centropages typicus	0.112	14	39.8 ± 6.1	22.1 ± 5.4	0.0120 ± 0.0031	0.256 ± 0.051	
Limnocalanus macrurus	0.19	18	55.1 ± 11.6	25.5 ± 4.7	0.0220 ± 0.0065	0.544 ± 0.108	
Pontella mediterranea	0.21	19	74.2 ± 24.6	44.0 ± 14.6	0.0112 ± 0.0025	0.469 ± 0.135	
Anomalocera patersoni	0.26	18	88.01 ± 8.9	57.1 ± 13.7	0.0095 ± 0.0014	0.532 ± 0.102	
Calanus helgolandicus	0.27	16	73.81 ± 8.3	45.8 ± 15.4	0.0150 ± 0.0050	0.629 ± 0.110	
Oncaea conifera	0.08	6		14.7 ± 2.4	0.0082 ± 0.0025	0.204 ± 0.021	
Corycaeus limbatus	0.07	4		11.3	0.0083	0.095	
Pseudocalanus elongatus	0.09	9	36.4 ± 6.1	21.2 ± 4.7	0.0068 ± 0.0007	0.142 ± 0.025	
Undinopsis similis	0.10	4		9.7 ± 3.5	0.0137 ± 0.0027	0.134 ± 0.013	
Pleuromamma abdominalis	0.24	10		25.0 ± 1.9	0.0147 ± 0.0002	0.386 ± 0.042	
Euchaeta media	0.24	5		18.3 ± 1.8	0.0121 ± 0.0013	0.220 ± 0.029	[29]
-"-	0.29	3		36.1 ± 2.2	0.0128 ± 0.0038	0.432 ± 0.047	
Euchirella messinensis	0.32	4	83.8 ± 22.0	41.5 ± 4.3	0.0153 ± 0.0008	0.708 ± 0.026	
-"-	0.39	3	116.0 ± 6.8	71.5 ± 4.5	0.0153 ± 0.0006	1.112 ± 0.105	
Anomalocera patersoni	0.38	5	102.9 ± 14.6	64.9 ± 8.3	0.0061 ± 0.0010	0.404 ± 0.108	
Oithona davisae	0.03	68	19.8 ± 4.2	10.1 ± 2.1	0.0074	0.075 ± 0.016	
Acartia tonsa	0.074	59	37.8 ± 9.6	24.1 ± 5.3	0.0076	0.185 ± 0.024	[22,28]
Calanus finmarchicus	0.30		75.6		0.013		
Acartia tonsa	0.083	55	44.6 ± 15	25.6 ± 10			
Acartia lilljeborgii	0.103	56	48.6 ± 11.7	23.2 ± 7.6			[19,21,58]
Temora turbinata	0.074	49	46.3 ± 5.3	25.3 ± 3.3			
Paracalanus parvus	0.066	30	40.7 ± 2.9	22.7 ± 2.0			
Temora turbinata	0.074		21.5 ± 5.5	10.3 ± 5.6			
Centropages furcatus	0.10		20.8 ± 1.7	11.5 ±1.6			[64]
Subeucalanus pileatus	0.205		45.3 ± 3.2	25.6 ± 2.5			
Pontella marplatensis	0.23		47.7 ± 17.2	24.3 ± 9.4			
Parvocalanus crassirostris	0.039		17		0.0034 ± 0.004	0.13 ± 0.01	[63]
Eurytemora affinis	0.077		34.2 ± 4.4	18.1 ± 10.2	0.0101 ± 0.001	0.21 ± 02	
Acartia hudsonica	0.075	14	38.7 ± 10.0				
Tortanus discaudatus	0.122	21	53.6 ± 5.7				[20]
Centropages hamatus	0.099	9	38.6 ± 2.8				
Temoralongicornis	0.059	4	26.2 ± 2.8				
Euchaeta elongata	0.41	8		31.4 ± 4.8			[65]
Euchaeta rimana	0.24	7		27.6 ± 3.2			
Paraeuchaeta elongata	0.40		120				[66]
Calanus pacificus	0.22	7	53 ± 7				
Bestiolina similis	0.054		26.3 ± 5.5				[67]

These data allowed us to examine the size-scaling of escape speeds stimulated by electric impulses at 20 °C over a large size range, and U_{max}, U_{min}, and U_{mean} all scaled approximately with prosome length to power of 3/4 (Figure 7).

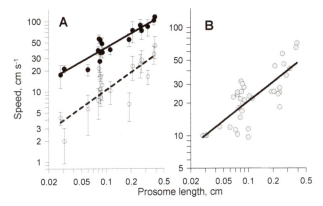

Figure 7. (**A**): Maximum (●) and minimum (○) species-specific instantaneous speed in a continuous sequence of kicks of the escape reaction stimulated by electrical impulses; $U_{max} = 194.9\ L^{0.66}$ ($R^2 = 0.87$) and $U_{min} = 70.0\ L^{0.83}$ ($R^2 = 0.70$). (**B**): Mean speed of escape reaction stimulated by various impulses, including predatory fish (data from Table 3); $U_{mean} = 82.0\ L^{0.60}$ ($R^2 = 0.62$).

3.2.2. Acceleration and Time Scale Features

Another important characteristic of the avoidance reaction is the acceleration of the body, which we calculated as $a = (U_{max} - U_{min})/t$, where U_{max} and U_{min} are the maximum and minimum speed during time of acceleration t. Body acceleration scales with size, approximately in the same way as jump speed (Figures 7A and 8A), while we did not find a significant effect of size on acceleration duration, nor on the duration of the power stroke. The total duration of a kick (D_{kick}), however, increased significantly with copepod size (Figure 8B).

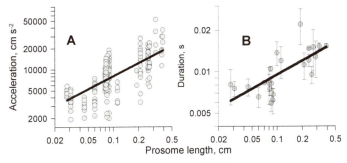

Figure 8. (**A**): Acceleration versus prosome length; $a \sim L^{0.62}$ ($R^2 = 0.50$). (**B**): Duration of kick; $D_{kick} = 0.021\ L^{0.34}$ ($R^2 = 0.51$).

3.2.3. Distance of Kicks

The number of kicks in a continuous series of escape reactions varies widely depending on the intensity and method of stimulation [21,22]. Usually, the maximum and mean speed of kicks decrease towards the end of the escape reaction due to the exhaustion of the energy resource. In addition to our old and new data, we were able to use only a few literature sources to analyze the escape movement of copepods (see Supplement Table S1). The distance covered during both the copepod stroke phase S_{st} and the entire kick phase S_{kick} scaled with prosome length as $L^{0.8}$ and $L^{0.88}$, respectively (Figure 9).

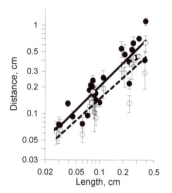

Figure 9. Distance covered during total kick phase (●) and stroke phase (○) approximated as S_{kick} = 1.55 $L^{0.88}$ (R^2 = 0.84) and S_{st} = 0.93 $L^{0.8}$ (R^2 = 0.82), respectively.

4. Force Estimation and Size Scaling

Forces of interest are those of drag and power stroke, and they can be determined in several ways. Drag can be directly measured by observing the sinking speed of models or immobilized specimens, or it can be measured indirectly by observing the non-propulsive deceleration of swimming specimens. The force production of beating appendages can be estimated from hydrodynamic models of the power stroke or from the equation of motion, observed velocity, and the acceleration of swimming specimens. The force production can also be directly estimated by measuring the force of hydrodynamically scaled physical models subject to a known water velocity, or it can be measured by a force sensor to which animals are attached.

4.1. Force Production in Copepods Tethered to Force Sensor

Comprehensive studies of the force production of copepods during cruising and jump reactions were performed using a semiconductor cantilever sensor [10,30–32,60,68] (Figure 10). The sensitivity of the sensors was sufficient to measure the force produced by the small cephalic appendages of the calanoid copepods *P. parvus* with a prosome length of 0.62 mm.

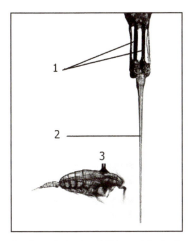

Figure 10. Force sensor (**1** and **2**: frontal view) and *Calanus helgolandicus* (*euxinus*) female (**3**: lateral view) attached to the end of the glass rod. **1**—four semiconductor tensoresistors of 2 × 0.2 × 0.05 mm, pairwise connected in one plane according to the scheme of the Wheatstone bridge; **2**—glass rod of 4 mm length in the case of measuring the force production of copepods during the escape reaction and 8–10 mm in the case of routine locomotion.

The results are shown in Figure 11 and Table 4. Figure 11A shows the integral average (defined as the area of pulse strength divided by pulse duration) of the force production by cephalic appendages during the cruise movement ($R_{p,cr,att}$) for eight species of the Black Sea and Mediterranean copepods with a prosome lengths (L) from 0.062 to 0.28 mm. Despite significant differences in the kinematics of the cephalic appendages in different species, the variation of $R_{p,cr,att}$ with L showed a high degree of correlation ($R^2 = 0.91$) approximated by the power-law (Figure 11A):

$$R_{p,cr,att} = 3.7\, L^{2.03}. \tag{1}$$

The same high correlation with the prosome lengths ($R^2 = 0.89$) was established for the average traction force ($R_{p,esc,att}$) of thoracic legs during escape reactions (Figure 11B):

$$R_{p,esc,att} = 384\, L^{2.2}, \tag{2}$$

as well as for the maximum instantaneous force during escape locomotion (Figure 11C; Table 3).

The average ratio of forces produced during escape and cruising locomotion has been seen to be about 100 (Table 4). This is much more than the ratio of forces during jumping and the displacement of higher aquatic and terrestrial animals, reaching only about 40 [69]. Of fundamental importance, Equations (1) and (2) show that the force production of both types of locomotion depends on the square of body size. This is consistent with M. Rubner's "surface rule", which states that in morphologically similar animals, the force available to them is proportional to the sectional area of the muscles or the square of the linear dimensions of the body [70]. Below, we consider the extent to which the length-square rule of the thrust force revealed on the attached copepods is confirmed by the kinematics and dynamics of their free swimming.

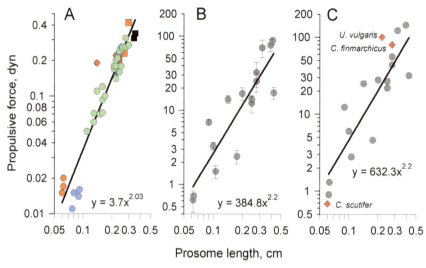

Figure 11. Propulsive force created by the Black and Mediterranean Seas copepods attached to semiconductor force sensor at 21 ± 2 °C (Table 3). (**A**): Mean resulting force of cephalic appendages in *Paracalanus parvus* (●), *Pseudocalanus elongatus* (●), *Calanus helgolandicus* (●), *Phaenna spinifera* (●), *Pontella mediterranea* (♦), *Pleuromamma abdominalis* (♦), *Euchaeta marina* (■), and *Euchirella messinensis* (■) (from Svetlichny 1993a). (**B**): mean tractive force of swimming thoracic legs during the escape reaction in *Paracalanus parvus*, *Acartia clausi*, *Calanus helgolandicus*, *Pontella mediterranea*, *Undinopsis similis*, *Scolecithrix Bradyi*, *Nannocalanus minor*, *Pleuromamma abdominalis*, *Eucalanus attenuates*, and *Euchirella messinensis*. (**C**): Maximum values of the species from (**B**) and species from literature data on *Cyclops scutifer* [33], *Undinula vulgaris* [34], and *Calanus finmarchicus* [61].

Table 4. Propulsive forces created by cephalic limbs during cruising and by thoracic legs at escape reaction in copepods attached to force sensor. The number of individuals is shown in parenthesis.

Species	L_{pr}, cm	Propulsion Force, Dyn			Source
		Cruising	Escape Reaction		
		Mean Integrated	Mean Integrated	Maximum Force	
Paracalanus parvus	0.062	0.018 ± 0.004 (2)	0.62 ± 0.2 (7)	0.9	
Acartia clausi	0.063		0.7 (2)	1.3	
-"-	0.106		1.5 ± 0.3 (4)	2.8	
Pseudocalanus elongatus	0.085	0.014 ± 0.0022 (4)			
Calanus helgolandicus	0.18	0.081 ± 0.02 (8)	2.4 ± 0.5 (4)	4.6	
-"-	0.25	0.019 ± 0.03 (7)	12.5 ± 3.3 (14)	22	
-"-	0.28	0.23 ± 0.04 (7)	24.8 ± 7.1 (8)	44	
-"-	0.28	0.28 ± 0.03 (3)	32.4 ± 11.9 (12)	56	
Pontella mediterranea	0.2	0.22 ± 0.013 (4)	16.9 ± 3.4 (6)	28	[32,60]
Undinopsis similis	0.1		3.3 ± 0.4 (4)	6	
Scolecithrix Bradyi	0.09		6.9 ± 0.6 (4)	12.3	
Phaenna spinifera	0.14	0.19 (1)			

Table 4. *Cont.*

Species	L_{pr}, cm	Propulsion Force, Dyn			Source
		Cruising	Escape Reaction		
		Mean Integrated	Mean Integrated	Maximum Force	
Nannocalanus minor	0.14		14.1 ± 1.7 (3)	25	
Pleuromamma abdominalis	0.25	0.22 ± 0.04 (4)	14.2 ± 1 (11)	27	
Eucalanus attenuatus	0.42		17.3 ± 3.2 (6)	32	
Euchaeta marina	0.32	0.37 ± 0.08 (2)			
Euchirella messinensis	0.32	0.34 (1)	70 ± 18.4 (4)	123	
-"-	0.39		76 ± 14.4 (3)	145	
-"-	0.41		87 ± 9.3 (4)	159	
Cyclops scutifer	0.06			0.68	[33]
Undinula vulgaris	0.22			125	[34]
Calanus finmarchicus	0.28			80	[61]

4.2. Drag on Falling Models and Specimens

The first task in the study of force production in freely moving copepods was to determine the drag on the body. Often, results for geometrically simple bodies are used as an approximation: a sphere (e.g., [22]) or an ellipsoid of revolution simulating the body of calanoid copepods *Paracalanus* and *Centropages* without protruding organs [71]. Here, we estimated drag coefficients, C_d, on carved wooden scale models that passively descended in a viscous fluid with different body orientations and antennae positions [72,73] (Figure 12). The hydrodynamic drag coefficient C_d was determined from the defining relation:

$$C_d = 2R_d/\rho S U^2, \qquad (3)$$

where the drag force R_d equals the submerged body weight (dyn), ρ (g cm^{-3}) is the density of the liquid, S (cm^2) is the sectional area (taken to be the area of a circle with a diameter d equal to the width of the prosome), and U is the observed sinking speed. C_d depends on the Reynolds number, Re = $\rho L U/\mu$, where μ denotes the dynamic viscosity. To compensate for the enlarged scale in these experiments, viscosity was adjusted by using glycerin–water mixtures (hydrodynamic scaling).

Later, the same principle was applied to immobilized individuals of 17 species of copepods [74]. After immobilization, the copepods were weighed in water on a modified Salvioni balance to determine their submerged body weight, and the rate of passive sinking was determined. To expand the range of Reynolds numbers, microparticles of lead were inserted into the body cavity. The drag coefficients of the body calculated from Equation (3) on the basis of the weight and speed of passive sinking are presented in Figure 12.

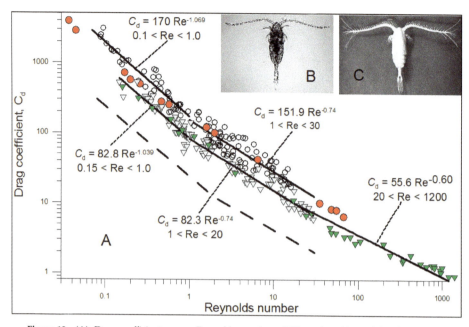

Figure 12. (**A**): Drag coefficient versus Reynolds number of 17 species of immobilized copepods (from [74]) when moving in water with open antennae (○) and antennae pressed to the body (▽). Red circles and green triangles indicate the C_d of enlarged models when moving with spread and pressed antennas, respectively (from [72,73]). The dashed line shows C_d of the sphere [75]. (**B**,**C**): Photos of immobilized calanoid copepod *Paracalanus parvus* and its enlarged (~1:100) model, respectively.

In Figure 12, two groups of data are distinguished: the case of movement with spread antennas, which is typical for slow cruise swimming, and the case of movement with folded antennas, which is typical for jumping movement. In general, the data turned out to be close to those obtained on enlarged models (Figure 12). To simplify the relationship, $C_d \sim f\{Re\}$ was approximated in each range of the Re scale by the relation [76]:

$$C_d = c\,Re^{-n}, \tag{4}$$

where c is the hydrodynamic shape factor and $Re = d\,U/\nu$, where ν is the kinematic viscosity, cm^2 s^{-1}, and d is body diameter (cm) corresponding to the largest width of the prosome. The estimated coefficients c for the different Re ranges (0.1–30.0 and 0.15–1200 Re for cruising and jumping, respectively) are shown in the correlation equations in Figure 12. Below, we use the experimentally determined drag coefficients to estimate force production from observed swimming speed and acceleration.

4.3. Detailed Analytical Model of Cruising Locomotion

At steady rectilinear translational motion, the drag of the body R_d equals the resulting propulsive force R_p created by the limbs in a time-averaged sense:

$$R_d = R_p, \tag{5}$$

If we multiply Equation (5) by the body velocity, the power $N_d = R_d\,U$ is the energy dissipation by drag, which equals the power effectively transferred to maintain the motion: $N_p = R_p\,U_{legs}$. However, the power actually expended by the limbs is much greater: $N_{legs} \gg N_d = N_p$, where N_{legs} is the total power of action of all cephalic limbs: second antennae, mandibles, maxillae, and maxillipeds (in *Calanus*, for example, with type of feeding; Figure 1A), because not all expended power by legs results in thrust.

To determine the power actually expended by the limbs, detailed measurements of the force and speed of individual cephalic limbs of attached cruising *Calanus helgolandicus* were carried out [10]. By determining the individual force production by second antennae, mandibles, maxillae, and maxillipeds after removing all other pairs of head limbs, it was found that the sum of these individual force productions added up to three times the force production of an intact specimen. Hence, the total power of all beating legs in this species can be estimated from the empirical relation:

$$N_{legs,att} = 3 R_p U_a, \qquad (6)$$

$$U_a = 2 \pi (\alpha/180) F l_a, \qquad (7)$$

where U_a is the circular speed of the second antenna relative to body, F denotes frequency of beat, l_a is the second antenna length measured from the point of attachment to the body to the middle of the length of the end bristles (Figure 13), and α is the angular amplitude of the legs rotation that varies near 50° for feeding current feeders like *P. elongatus*, amounts to 80–90° for cruising feeders, like *C. helgolandicus*, and amounts to 100–120° for pontellids species (our personal observations based on high speed video; see Figure 1).

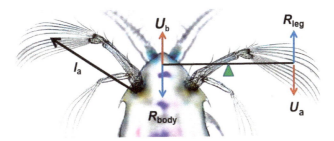

Figure 13. Second antennas of *Pontella mediterranea*. The black arrow on the left shows the length of the antennae l_a; other arrows show the forces and speeds of the body and legs. The magnitudes of velocities U_b (= U) and U_a determine the location of a simulated rowlock (green triangle) of an oar model.

In free swimming copepods, however, the beating legs act on water with the effective velocity of (U_a-U), and, by taking into account the empirical value k (possibly different from 3) of the hydrodynamic efficiency of locomotion, the total power of all beating legs of the free cruising copepods can be determined as:

$$N_{tot} = R_d U + k R_p (U_a - U), \qquad (8)$$

The first term in Equation (8) is the power of thrust transferred to the body for it to overcome the drag (i.e., the useful thrust power), while the second empirical term represents the extra power dissipated by the moving limbs, a quantity that is not useful for propulsion. The drag force on the body is calculated based on the average speeds for each of the studied species (Table 2) from the usual equation of drag expressed in terms of an empirical drag coefficient C_d (recall Equation (3)):

$$R_d = \frac{1}{2} C_d \rho S_{body} U^2, \qquad (9)$$

Taking $S = \pi d^2/4$ and C_d from Equation (4) with n = 0.74 for 1 < Re < 30 (see Figure 12), we obtain:

$$R_d = 59.7 \, v^{0.74} \, \rho_w \, d^{1.26} \, U^{1.26}, \qquad (10)$$

Using the data of Table 2 for cruising copepods led to the scaling $R_d = 11.5\ L^{2.82}$ ($R^2 = 0.86$) (Figure 14), which differed from the scaling $R_{p,att} \sim L^{2.03}$ for attached copepods (Figure 11).

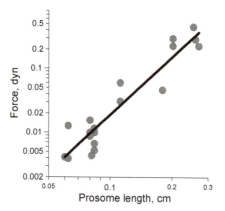

Figure 14. Drag force versus prosome length calculated for free cruise-swimming copepods, ($R_d = 11.5\ L^{2.82}$ ($R^2 = 0.86$). All points are average values.

In the smallest tethered calanoid copepods *Paracalanus parvus* and *Pseudocalanus elongatus*, $R_{p,att}$ was significantly ($p < 0.001$) 2.5 times higher than R_d (0.017 ± 0.005 and 0.0068 ± 0.004 dyn, respectively), whereas in the largest species, there was no difference because of different values of the empirical factor k in Equation (8), as shown in [10] and seen from Figure 14, probably due to the higher hydrodynamic efficiency of the paddle locomotion at higher Reynolds numbers.

It has previously been shown that the flow field around tethered copepods differs from that around a grazing free-swimming animal [20,27,56,77]. However, the difference in the scaling of force production output and available force to overcome body drag may also be due to a change in the hydrodynamic efficiency of the type of locomotion (coefficient k in Equation (8)). For this reason, the muscle force realized by attached individuals approximately scales as $R_{p,att} \sim L^2$, but it scales as $R_d \sim L^3$ in freely cruising copepods (Figure 14). Taking into account that the same species were used in our experiments with attached and free copepods, we could test this hypothesis by calculating the propulsive force R_p of freely moving individuals as the drag force on beating limbs using the following equation:

$$R_p = \frac{1}{2} C_{d,leg,att}\ \rho_w\ S_{leg}\ (U_a - U)^2, \tag{11}$$

where U_a is the circular speed of the second antenna (Equation (7)), S_{leg} is the cross sectional area of legs, and $C_{d,leg,att}$ is the drag coefficient of attached individuals calculated as:

$$C_{d,leg,att} = 2\ R_{p,cr,att}/\rho_w\ S_{leg}\ U_a^{\ 2}, \tag{12}$$

From all measurements, we found the correlations $C_{d,leg,att} = 34.5\ Re_l^{-0.88}$ ($R^2 = 0.91$), where $Re_l = U_a\ l_a/\nu$ and $R_p = 5.46\ L^{2.36}$ (Figure 15). This may indicate that the propulsive force of the limbs, directly measured in tethered copepods or predicted for free-swimming individuals, is more consistent with the scale L^2.

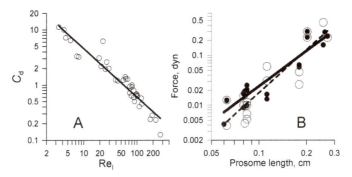

Figure 15. (**A**): Coefficient of hydrodynamic resistance $C_{d,leg,att}$ (○) from Equation (12) for cephalic limbs of 8 copepod species attached to a force sensor. (**B**): Propulsion force R_p (●) calculated from experimental data for the same free cruise-swimming copepods based on $C_{d,leg,att}$ and, for comparison, the values of drag force R_d (○, dotted line) calculated by Equation (10).

Next, we calculated the power required to overcome body drag and resistance of cephalic limbs' actions, the two terms $N_d = R_d U$ and $N_p = k(U_a - U) R_p$ in Equation (8). The results in Figure 16 indicate that power that is sufficient overcome body drag scales as $N_d \sim L^{4.1}$, while for limbs, it scales as $N_p \sim L^{3.1}$ or as $\sim M^{1.0}$, where M denotes body mass. A similar regression coefficient ($L^{3.04}$) was obtained when calculating the power of attached cruising copepods using equation $N_{p,att} = k U_a R_{p,att}$.

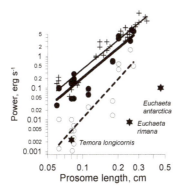

Figure 16. Calculated power versus prosome length required for free cruise-swimming copepods to overcome body drag (○, thin dotted line, $N_d = 137.7 L^{4.09}$) and to move cephalic limbs (●, solid line, $N_p = 266.07 L^{3.1}$), as well as the results for attached individuals (+, thin line, $N_{p,att} = 436.57 L^{3.04}$). Black asterisks indicate literature data for *Temora longicornis* [17]; *Euchaeta rimana* [26], and *Euchaeta Antarctica* [27].

To calculate N_p, we used the empirical value from Equation (6), $k = 3$, for *C. helgolandicus* [10]. However, when we took into account the difference in the type of cephalic appendages action and the efficiency of locomotion of small cruising feeder copepods compared to large cruising feeders and especially pontellids, in which the cephalic appendages do not oppose each other during the creation of propulsive force, the slope of the regression line became less than 3.0. In other words, the scaling $R_p \sim L^{3.0}$ can lead to an underestimation of the power consumption of small species and an overestimation in large ones. Such a correction corresponds to the scaling of the energy potential of animals [78] whose biological power is usually proportional to $M^{0.67-1.0}$.

4.4. Analytical Model of Escape Reaction

One way to obtain estimates of forces and energy change during an escape jump from measured kinematics is to use the equation of motion of body mass M during acceleration dU/dt due to propulsive force R_p and opposing drag R_d:

$$M\, dU/dt + R_d = R_p, \tag{13}$$

First, for non-propulsive deceleration, Equation (13) provides an estimate of the drag force as a function of velocity:

$$R_d = -M\, dU/dt, \tag{14}$$

However, such estimates prove to be quite inaccurate because they depend on the numerical discretization of the time derivative of second order of position. Therefore, it is more accurate to assume the validity of measured relations for the drag coefficient of sinking specimens (Figure 12) and calculate the drag force from the usual relation Equation (9):

$$R_d = \frac{1}{2}\, C_d\, \rho_w\, (\pi/4)\, d^2\, U^2, \tag{15}$$

where $C_d = 55.6\, Re^{-0.60}$ for the range of $10 < Re < 1200$ (Figure 12), which corresponds to our studied copepods with a body width of $d = 0.013$–0.13 cm and a mean speed $U = 10$–100 cm s^{-1} at constant temperature of 20 °C (Table 3, Supplement Table S1).

The values of R_d calculated by Equation (15) using the average speed of cyclopoid and calanoid copepods during the stroke phase of escape reactions increased on the average from 0.1 dyn in small oithonids to 30 dyn in the largest calanoid copepods (Figure 17A) according to the scaling $R_d \sim L_p^{2.15}$. Using this relation and observed accelerations in Equation (13) led to the scaling $R_p \sim L_p^{2.55}$. A similar procedure for attached calanoid copepods gave the close scaling $R_{p\,att} \sim L_p^{2.37}$ (Figure 17B).

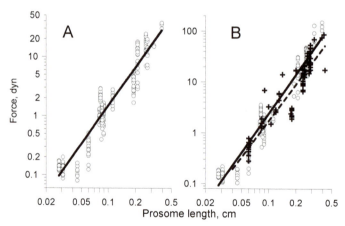

Figure 17. (**A**): Drag force; $R_d = 211.6\, L^{2.15}$ ($R^2 = 0.91$, $N = 241$). (**B**): Propulsive force, calculated (○) for free swimming copepods and directly measured (+) in attached individuals during an escape reaction; $R_p = 838.2\, L^{2.55}$ ($R^2 = 0.94$) and $R_{p,att} = 408.6\, L^{2.37}$ ($R^2 = 0.82$, $N = 88$).

Multiplying the equation of motion (Equation (13)) by the velocity of the body during kick stroke phases and integrating it over the time of acceleration during which the velocity increases from U_{min} to U_{max}, the energy expended (ΔE_{stroke}) is obtained as:

$$M\frac{1}{2}(U_{max}^2 - U_{min}^2)_1 + <U\, R_d\, \Delta t>_1 = <U\, R_p\, \Delta t>_1 \equiv \Delta E_{stroke}, \tag{16}$$

where $<>_1$ signifies an integrated quantity over time interval Δt_1.

Following the power stroke phase, the limbs are retuned back during time Δt_2, while the velocity decreases from U_{max} in the end of stroke phase back to U_{min}, for which the energy balance gives:

$$M \frac{1}{2}(U_{max}^2 - U_{min}^2)_2 - <U\,R_d\,\Delta t>_2 = \Delta E_{limb,back}, \qquad (17)$$

which merely shows that deceleration is caused by body and limb drag.

Including the so-called 'energy-leg-back' contribution, the total energy expended by limbs during all kick stroke phases is $E_{sum} = \Delta E_{stroke} + \Delta E_{limb,back}$, or:

$$E_{sum} = M \frac{1}{2}(U_{max}^2 - U_{min}^2)_1 + <U\,R_d\,\Delta t>_1 + M \frac{1}{2}(U_{max}^2 - U_{min}^2)_2 - <U\,R_d\,\Delta t>_2, \qquad (18)$$

Using Equation (18), the mean power of an escape kick N_{esc}, defined as $N_{esc} = E_{sum}/D_{kick}$ (where the duration is $D_{kick} = \Delta t_1 + \Delta t_2$) was calculated to vary in the range from 1 to 4000 erg s^{-1} following the scaling $N_{esc} \sim L^{3.05}$ (Figure 18). This result turned out to be very close to the power of attached calanoids that scale as $N_{esc,att} \sim L^{2.99}$, which was calculated as $N_{esc,att} = R_{p,att}\,U_{leg}$, where $U_{leg} = 2\,\pi\,(\alpha/180)\,F\,h_a$, $\alpha = 145°$, and $h_a = 0.75\,l_a$ according [23]. In both cases, N_{esc} was seen to scale linearly with body mass M.

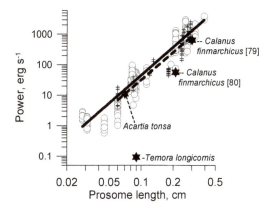

Figure 18. Power of kick N_{esc} during the escape reaction of free swimming ($\bigcirc$) and attached (+) copepods. The power regressions were $N_{esc} = 51000\,L^{3.05}$ ($R^2 = 0.91$, $N = 190$) and $N_{esc,att} = 31400L^{2.99}$ ($R^2 = 0.82$, $N = 71$), respectively. Black asterisks indicate literature data for *Temora longicornis* [25]; *Acartia tonsa* and *Calanus finmarchicus* from [79], and *C. finmarchicus* according [80].

5. Discussion and Conclusions

The cruising speeds of calanoid copepods vary widely depending on the type of feeding and the associated mechanism of creating propulsive force, on the body density, and on the water temperature. The density of the body is significantly higher than the density of water [45]. In this regard, the speed of passive sinking can distort the real speeds that are provided by the movement of the limbs. For many calanoid copepods, the available cruising speed is only two-to-three times higher than the speed of gravitational sinking [45,81,82]. For example, in females of *C. helgolandicus* at 20 °C, sinking speed can reach 0.8 cm s^{-1} [83]. Therefore, at the maximum swimming speed available to them (see Table 1), their speed changes 2.5 times from 1.4 to 3 cm s^{-1}, depending on the direction of movement being down or up. Temperature affects speed through changes in the viscosity and density of water [84], but it changes to a greater extent due to changes in the rate of muscle contraction. The rate of many biological systems, including planktonic crustaceans [35], varies in proportion to the temperature coefficient $Q_{10} = 2$, i.e., an increase of a factor 2 when the temperature increases by 10 °C. This has been confirmed in experiments examining the temperature response of limb beat frequency and the swimming speed of copepods [54,82,85]. Therefore, we used video recordings of horizontal cruise swimming calanoid copepods from the Black, Marmara, and Baltic Seas at the same temperature 20 °C, as well as literature data for swimming copepods at similar temperatures.

5.1. Scaling of Kinematic and Mechanical Parameters of Cruising

The average cruising speed and cephalic limb beat frequency scaled as $U = 13.4\ L^{1.4}$ (Table 2) and $F = 16.0\ L^{-0.53}$, respectively. According to the reviews [86,87] that investigated the scale laws of mechanics and kinematics of "biological motors" of different systematic groups, such empirical slopes of U and F correspond to cyclic motors with mass $M > \sim 0.4$ mg (fruit fly size and above) whose maximum force output scales as $R \sim L^{3.0}$ or $R \sim M^{1.0}$.

In our analysis of free swimming cruising copepods, we found a scaling of the body drag force $R_d \sim L_{pr}^{2.82}$ similar to that of cyclic motors (Table 5), whereas in tethered copepods, the measured force production scaled as $R_{p\ cr,att} \sim L_{pr}^{2.06}$ or $M^{0.69}$. According to [86], animals whose maximum force output scales as $M^{0.67}$ correspond to a group of steady translational (i.e., linear) motors. However, Marden [86] noted that: "there are potentially many force outputs by translational motors .... that fall between the two fundamental scaling relationships ... " $R \sim L^{2.0}$ and $R \sim L^{3.0}$.

Table 5. Exponents m in scaling relations versus body length, L^m.

Quantity and Condition	Cruising				Escape Jump			
	Free Swimming		Attached Locomotion		Free Swimming		Attached Locomotion	
	m	Figure	m	Figure	m	Figure	m	Figure
Body speed, U	1.4	Figure 4			0.7	Figure 7		
Drag force, R_d	2.82	Figure 14			2.15	Figure 17A		
Propulsive force, R_p	2.36	Figure 15	2.06	Figure 11A	2.55	Figure 17B	2.2	Figure 11B
Power	3.1	Figure 16	3.04	Figure 16	3.05	Figure 18	2.94	Figure 18

Note that the above difference in scaling of R_d and $R_{p,cr,att}$ revealed by us was mainly due to smaller values of R_d in small species (see Figure 14), the magnitude of which can be illustrated as follows. In order for the predicted R_d of smallest free swimming *P. parvus* to increase to the level of $R_{p,cr,att}$ in the attached individuals of this species, their average speed should be two times higher than our measured speeds (see Table 1). Hence, scaling according to $L^{2.0}$ may be the best estimate for all sizes. The total cruising power N_{cr} of copepods in the size range $0.06 < L < 0.3$ cm, calculated on the basis of the force and speed of the cephalic appendages, varied on average from 0.05 to 5 erg s^{-1} (or from 0.05 to 5×10^{-7} W) in proportion to $L^{3.1}$ or $\sim M^{1.0}$ (Figure 16). This is consistent with the scaling of metabolic energy available for the long-term cruising of animals, which usually scales as $M^{0.67–1.0}$ [78], while the net power needed to move the body, calculated based on body drag N_d and speed, has an excessively high exponent $L^{4.1}$ or $M^{1.4}$ (Figure 16). According to our estimation, the efficiency of locomotion defined as N_d/N_{sum} changed, on average, from 5% in *P. parvus* up to 20% in pontellids.

Few other studies have dealt with the mechanical power of cruise swimming copepods, and all of these have calculated the rate of energy dissipation in the liquid volume due to the movement of the limbs of a cruising copepod. For only one species, an adult female *Temora longicornis* [17], the power (2.3×10^{-10} W) was close to our estimated power to overcome the body drag in copepods of the same size ($L \sim 0.08$ cm, about 3×10^{-10} W, Figure 16). In two other similar studies, the energy dissipation by *Euchaeta* rimana [26], and especially *Euchaeta antarctica* [27], turned out to be almost two orders of magnitude smaller than for copepods of the similar size from our experiments. The discrepancy can be partially explained by the fact that this very large Antarctic copepod swam in cold water (0 °C) at a speed (1.5 cm s^{-1}) that was approximately three times lower than the expected speed at 20 °C in a copepod of the same size (Figure 4A). Similarly, the speed of subtropical *E. rimana* at 20 °C (0.7 cm s^{-1}) [57] was three times lower than that of *C. helgolandicus* of the same size.

5.2. Scaling of Kinematic and Mechanical Parameters of Escape Reaction

The escape reaction for all copepods is carried out by a simple sequence of strokes with morphologically similar thoracic swimming legs [37] and, apparently, with similar efficiency. Therefore, the predicted correlations of R_d and R_p for free swimming and R_p measured in tethered copepods during escape reaction were more consistent with each other than in the case of cruising (Table 3).

The observed scaling of escape speeds with body size, $U_{mean} \sim L^{0.7}$ and $U_{max} \sim L^{0.66}$, as well as drag and force production (Table 3), are more consistent with the translational motors whose maximum force output scales as $L^{m<3.0}$ [86]. Indeed, the measured propulsive force of copepods attached to the force sensor scaled as $R_{p,att} \sim L^{2.15}$, and the calculated forces of free escapes scaled as $R_{d,free} \sim L_p^{2.36}$ and $R_{p,free} \sim L^{2.55}$. The average values of $R_{p,att}$ for the smallest calanoid copepod P. parvus (0.62 ± 0.2 dyn), as well as for the largest E. messinensis (87 ± 9 dyn), did not differ significantly from the calculated values of $R_{p,free}$ (Figure 17B).

Nevertheless, the total power of free copepods during the escape reaction turned out to scale as $N_{esc} \sim L^{3.06}$ and for the attached as $N_{esc} \sim L^{2.94}$. Thus, the total power of both free and attached copepods during the escape reaction turned out to scale as $L^{3.0}$. This trend in N_{esc} was confirmed by the results of calculations by Jiang and Kiørboe [79], who estimated the maximum values of mechanical power for Acartia tonsa (0.069 cm prosome length) and Calanus finmarchicus with a prosome length of 0.3 cm as 1.1×10^{-6} and 6.3×10^{-5} W, respectively. Muphy et al. [80] determined the value of maximum power delivered to the fluid by the swimming legs of C. finmarchicus (L = 0.21 cm) to be 5.6×10^{-6} W. The maximum energy delivered by swimming appendages defined by Duren and Videler [25] in Temora longicornis (L = 0.09 cm) equaled 9.3×10^{-9} W. This was almost two orders less, probably due to the relatively low U_{max} of the studied individuals (10.8 cm s^{-1}) in comparison with the U_{max} of the escape reaction of this species stimulated by hydrodynamic stimuli (26.2 cm s^{-1}) [20].

5.3. Cost of Transport during Cruising and Jumping

In general, the propulsive force and the power created by the swimming limbs are two and three, respectively, orders of magnitude higher than the force and power created by the head appendages. However, it is more correct to assess the differences between the two types of swimming of the copepods by the energy costs of transport (C_t) [78], defined as the energy consumption per unit of body mass and distance travelled (S): $C_t = E/M\,S = N/M\,U$ (cal g^{-1} km^{-1}).

For large calanoid copepods (L = 0.2–0.3 cm), the average mechanical cost of transport moving by unsteady jumps (C_{tm} = 45.2 ± 15 cal g^{-1} km^{-1}) is seven times higher than by steady cruise swimming (C_{tm} = 6.7 ± 3.1 cal g^{-1} km^{-1}), while for small calanoid copepods (L = 0.06 cm), it is only about three times higher (C_{tm} = 74.4 ± 24 and C_{tm} = 25.1 ± 16 cal g^{-1} km^{-1}, respectively) and the ratio is even less in the smallest copepods (Figure 19A). Thus, for large copepods, the cost of transportation is much higher for swimming-by-jumping than for cruise swimming, while for small ones, the difference is not so large. There are advantages to swimming-by-jumping, the first being hydrodynamic stealth: swimming-by-jumping creates only a relatively small fluid disturbance and, thus, is less susceptible to rheotactic predators than copepods that cruise steadily [28]. This may explain why only small copepods (cyclopoids) swim by jumps, while larger copepods are cruise swimmers.

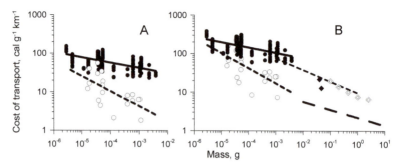

Figure 19. (**A**): Maximum mechanical cost of transport (C_{tm}) during escape reaction (•) and cruising (○). (**B**): Metabolic cost of transport (C_{tb}) and values for swimming fish (long dashed line) and flight of insects (short dashed line) [78]; escape reaction of shrimp (unshaded diamond) [88] and *Euphausia* (black diamond) [89].

The biological cost of transport C_{tb} is due not only to the mechanical efficiency of locomotion but also to the efficiency of muscle contraction. The theoretical maximum efficiency of muscle contraction efficiency is 0.5 [90]. However, with prolonged cruise work, the maximum coefficient of mechano-muscular efficiency of aerobic muscles does not exceed 0.25. With short-term muscle action during the escape reaction, it can increase to 0.4 [91]. To compare our measurements with observations for other species recorded in the literature, we multiplied our estimates of the mechanic costs of transportation by factors of 4 and 2.5 for cruising and escape jumping, respectively (Figure 19B). Transportation costs for escape jumps were found to be in line with those for other arthropods [23,24], and cruise swimming was found to be consistent with swimming costs in fish (not startle responses).

Supplementary Materials: The following is available online at http://www.mdpi.com/2311-5521/5/2/68/s1, Table S1: Kinematic parameters of the escape reaction in calanoid and cyclopoid copepods at 20–22 °C.

Author Contributions: L.S. conceived, designed, and performed the experiments; L.S., P.S.L., and T.K. analyzed the data; all authors contributed to writing and have approved the final version of the manuscript. All authors have read and agreed to the published version of the manuscript.

Funding: The Center for Ocean Life is supported by the Villum Foundation. We further acknowledge support from the Danish Council for independent Research (7014-00033B).

Acknowledgments: This work was supported by the projects of the NASU (grant number 0114U002041). The experimental part of this study was carried during 1984–2016 in the Department of animal physiology of Institute of Biology of the Southern Seas, Sevastopol, Ukraine, Faculty of Aquatic Sciences of Istanbul University (2007–2019), Turkey and in the SYKE Marine Research Center (2019), Finland.

Conflicts of Interest: The authors declare no conflicts of interest.

References

1. Huys, R.; Boxshall, G.A. The Orders of Copepods. In *Copepod Evolution*; The Ray Society: London, UK, 1991; p. 159.
2. Ho, J.-S. Copepod phylogeny: A reconsideration of Huys & Boxshall's 'parsimony versus homology'. *Hydrobiologia* **1994**, *292–293*, 31–39.
3. Cannon, H.G. On the feeding mechanism of the copepods *Calanus finmarchicus* and *Diaptomus gracialis*. *Br. J. Exp. Biol.* **1928**, *6*, 131–144.
4. Storch, O.; Pfisterer, O. Der Fangapparat von Diaptomus. *J. Comp. Physiol. A* **1925**, *3*, 330–376. [CrossRef]
5. Lowndes, A.G. The swimming and feeding of certain calanoid copepods. *Proc. Zool. Soc. Lond.* **1935**, *3*, 687–715. [CrossRef]
6. Gauld, D.T. The swimming and feeding of planktonic copepods. In *Some Contemporary Studies in Marine Science*; Barnes, H., Ed.; Allen and Unwin: London, UK, 1966; pp. 313–333.

7. Petipa, T.S. Idem. In *Trophodynamics of Copepods in Marine Plankton Communities*; "Nauk. Dumka" Press: Kiev, Ukraine, 1981; p. 241.
8. Koehl, M.A.R.; Strickler, J.R. Copepod feeding currents: Food capture at low Reynolds number. *Limnol. Oceanogr.* **1981**, *26*, 1061–1073. [CrossRef]
9. Price, H.J.; Paffenhofer, G.-A. Capture of small cells by the copepod Eucalanus elongatus. *Limnol. Oceanogr.* **1986**, *31*, 189–194. [CrossRef]
10. Svetlichny, L.S. Filming, tensometry and energy estimation of swimming by mouth appendages in *Calanus helgolandicus* (Crustacea, Copepoda). *Zool. J.* **1991**, *70*, 23–29. (In Russian)
11. Rosenberg, G.G. Filmed observations of filter feeding in the marine plankton copepod *Acartia clausii*. *Lirnnol. Oceanogr.* **1980**, *25*, 738–742.
12. Svetlichny, L.S. Locomotor function of mouth appendages in copepods: Its kinematics. *Ekol. Morya* **1993**, *44*, 84–91. (In Russian)
13. Borg, C.M.A.; Bruno, E.; Kiørboe, T. The Kinematics of Swimming and Relocation Jumps in Copepod Nauplii. *PLoS ONE* **2012**, *7*, e47486. [CrossRef]
14. Storch, O. Die Schwimmbewegung der Copepoden, auf Grund von Mikro-Zeitlupenaufnahmen analysiert. *Verh Dtsch Zool Ges* **1929**, *4*, 118–129.
15. Strickler, J.R.; Bal, A.K. Setae of the first antennae of the copepod *Cyclops scutifer* (Sars): Their structure and importance. *Proc. Natl. Acad. Sci. USA* **1973**, *70*, 2656–2659. [CrossRef] [PubMed]
16. Petipa, T.S. Methods of movement and food capture in *Calanus helgolandicus* (Claus). In *Biology and Distribution of the Plankton of the South Seas*; "Nauka" Press: Moscow, Russia, 1967; pp. 37–57.
17. Van Duren, L.A.; Stamhuis, E.J.; Videler, J.J. Copepod feeding currents: Flow patterns, filtration rates and energetics. *J. Exp. Biol.* **2003**, *206*, 255–267. [CrossRef] [PubMed]
18. Van Duren, L.A.; Videler, J.J. The trade-off between feeding, mate seeking and predator avoidance in copepods: Behavioural responses to chemical cues. *J. Plankt. Res.* **1996**, *18*, 805–818. [CrossRef]
19. Buskey, E.J.; Lenz, P.H.; Hartline, D.K. Escape behavior of planktonic copepods in response to hydrodynamic disturbances: High-speed video analysis. *Mar. Ecol. Prog. Ser.* **2002**, *235*, 135–146. [CrossRef]
20. Burdick, D.S.; Lenz, P.H.; Hartline, D.K. Escape strategies in co-occurring calanoid copepods. *Limnol. Oceanog.* **2007**, *52*, 2373–2385. [CrossRef]
21. Waggett, R.J.; Buskey, E.J. Escape reaction performance of myelinated and non-myelinated calanoid copepods. *J. Exp. Mar. Biol. Ecol.* **2008**, *361*, 111–118. [CrossRef]
22. Kiørboe, T.; Andersen, A.; Langlois, V.; Jakobsen, H.H. Unsteady motion: Escape jumps in copepods, their kinematics and energetics. *J. R. Soc. Interface* **2010**, *7*, 1591–1602. [CrossRef]
23. Morris, M.J.; Gust, G.; Torres, J.J. Propulsion efficiency and cost of transport for copepods: A hydromechanical model of crustacean swimming. *Mar. Biol.* **1985**, *86*, 283–295. [CrossRef]
24. Morris, M.J.; Kohlhage, K.; Gust, G. Mechanics and energetics of swimming in the small copepod Acanthocyclops robustus (Cyclopoida). *Mar. Biol.* **1990**, *107*, 83–91. [CrossRef]
25. Van Duren, L.A.; Videler, J.J. Escape from viscosity: The kinematics and hydrodynamics of copepod foraging and escape swimming. *J. Exp. Biol.* **2003**, *206*, 269–279. [CrossRef] [PubMed]
26. Yen, J.; Sanderson, B.; Strickler, J.R.; Okubo, A. Feeding currents and energy dissipation by Euchaeta rimana, a subtropical pelagic copepod. *Limnol. Oceanogr.* **1991**, *36*, 362–369. [CrossRef]
27. Catton, K.B.; Webster, D.R.; Brown, J.; Yen, J. Quantitative analysis of tethered and free-swimming copepodid flow fields. *J. Exp. Biol.* **2007**, *210*, 299–310. [CrossRef]
28. Jiang, H.; Kiørboe, T. The fluid dynamics of swimming by jumping in copepods. *J. R. Soc. Interface* **2011**, *8*, 1090–1103. [CrossRef]
29. Svetlichny, L.S. Morphology and functional parameters of body muscles of *Calanus helgolandicus* (Copepoda, Calanoida). *Zool. J.* **1988**, *67*, 23–30. (In Russian)
30. Svetlichny, L.S. Speed, force and energy expenditure in the movement of copepods. *Oceanology* **1987**, *27*, 497–502.
31. Svetlichny, L.S. Escape reaction in the copepod *Calanus helgolandicus*. *Zool. J.* **1986**, *65*, 506–515. (In Russian)
32. Svetlichny, L.S. Locomotor function of mouth appendages in copepods: Hydromechanical and energetic similarity. *Ekol. Morya* **1993**, *44*, 91–99. (In Russian)
33. Alcaraz, M.; Strickler, J.R. Locomotion in copepods: Pattern of movements and energetics of *Cyclops*. *Hydrobiologia* **1988**, *167*, 404–414.

34. Lenz, P.H.; Hartline, D.K. Reaction times and force production during escape behavior of a calanoid copepod, *Undinula vulgaris*. *Mar. Biol.* **1999**, *133*, 249–258. [CrossRef]
35. Lenz, P.H.; Hower, A.E.; Hartline, D.K. Temperature compensation in the escape response of a marine copepod, *Calanus finmarchicus* (Crustacea). *Biol. Bull.* **2005**, *209*, 75–85. [CrossRef] [PubMed]
36. Kiørboe, T.; Saiz, E.; Tiselius, P.; Andersen, K.H. Adaptive feeding behavior and functional responses in pelagic copepods. *Limnol. Oceanogr.* **2018**, *63*, 308–321. [CrossRef]
37. Lewis, A.G.; Johnson, C.; Allen, S.E. Calanoid copepod thoracic legs-surface area vs. body size and potential swimming ability, a comparison of eight species. *Crustaceana* **2010**, *83*, 695–713. [CrossRef]
38. Kurbatov, B.V.; Svetlichny, L.S. Kinematics and hydrodynamical resistance of *Calanus helgolandicus* (Claus) thoracic limbs. *Ekol. Morya* **1982**, *10*, 75–81. (Russian with English summary)
39. Rader, B.W. *Rhincalanus Cornutus* (Copepoda): Trunk Skeletomusculature. *Trans. Am. Microsc. Soc.* **1970**, *89*, 75–99. [CrossRef]
40. Boxshall, G.A. The comparative anatomy of two copepods, a predatory calanoid and a particle-feeding mormonilloid. *Phil. Trans. R. Soc. Lond.* **1985**, *311*, 303–377.
41. Pringle, J.W.S. *Insect Flight*; Cambridge University Press: Cambridge, UK, 1957; p. 132.
42. Kiørboe, T.; Jiang, H.; Colin, S.P. Danger of zooplankton feeding: The fluid signal generated by ambush-feeding copepods. *Proc. R. Soc. B Boil. Sci.* **2010**, *277*, 3229–3237. [CrossRef]
43. Svetlichny, L.; Larsen, P.S.; Kiørboe, T. Swim and fly: Escape strategy in neustonic and planktonic copepods. *J. Exp. Biol.* **2018**, *221*, jeb167262. [CrossRef]
44. Strickler, J.R. Swimming of planktonic Cyclops species (Copepoda, Crustacea): Pattern, movements and their control. In *Swimming and Flying in Nature*; Wu, T.T., Brokaw, C.J., Brennan, C., Eds.; Plenum Press: New York, NY, USA, 1975; pp. 599–613.
45. Mauchline, J. The biology of calanoid copepods. In *Advances in Marine Biology*; Blaxter, J.H.S., Southward, A.J., Tyler, P.A., Eds.; Academic Press: Cambridge, MA, USA, 1998; p. 710.
46. Borazjani, I.; Sotiropoulos, F.; Malkiel, E.; Katz, J. On the role of copepod antennae in the production of hydrodynamic force during hopping. *J. Exp. Biol.* **2010**, *213*, 3019–3035. [CrossRef]
47. Kiørboe, T. Mate finding, mating, and population dynamics in a planktonic copepod *Oithona davisae*: There are too few males. *Limnol. Oceanogr.* **2007**, *52*, 1511–1522. [CrossRef]
48. Kiørboe, T. Optimal swimming strategies in mate searching pelagic copepods. *Oecologia* **2008**, *155*, 179–192. [CrossRef]
49. Pavlova, E.V. *The Movement and Metabolism of Marine Planktonic Organisms*; Naukova Dumka: Kiev, Ukraine, 1987; p. 212. (In Russian)
50. Swift, M.C.; Fedorenko, A.Y. Some aspects of prey capture by Chaoborus larvae. *Limnol. Oceanogr.* **1975**, *20*, 418–426. [CrossRef]
51. Price, H.J.; Paffenhofer, G.-A.; Strickler, J.R. Modes of cell capture in calanoid copepods. *Limnol. Oceanogr.* **1983**, *28*, 116–123. [CrossRef]
52. Cowles, T.J.; Strickler, J.R. Characterization of feeding activity patterns in the planktonic copepod *Centropages typicus* Krøyer under various food conditions. *Limnol. Oceanogr.* **1983**, *2*, 106–115. [CrossRef]
53. Chen, M.R.; Hwang, J.S. The swimming behavior of the calanoid copepod *Calanus sinicus* under different food concentrations. *Zool. Stud.* **2018**, *57*, 13. [CrossRef]
54. Gill, C.W.; Crisp, D.I. The effect of size and temperature on the frequency of limb beat of *Temora longicornis* Miller (Crustacea: Copepoda). *J. Exp. Mar. Biol. Ecol.* **1985**, *86*, 185–196. [CrossRef]
55. Castel, J.; Veiga, J. Distribution and retention of the copepod *Eurytemora affinis hirundoides* in a turbid estuary. *Mar. Biol.* **1990**, *107*, 119–128. [CrossRef]
56. Bundy, M.H.; Paffenhöfer, G.A. Analysis of flow fields associated with freely swimming calanoid copepods. *Mar. Ecol. Progr. Ser.* **1996**, *133*, 99–113. [CrossRef]
57. Yen, J. Directionality and swimming speeds in predator-prey and male–female interactions of Euchaeta rimana, a subtropical marine copepod. *Bull. Mar. Sci.* **1988**, *43*, 395–403.
58. Buskey, E.J.; Hartline, D.K. High-speed video analysis of the escape responses of the copepod Acartia tonsa to shadows. *Biol. Bull.* **2003**, *240*, 28–37. [CrossRef]
59. Hartline, D.K.; Buskey, E.J.; Lenz, P.H. Rapid jumps and bioluminescence elicited by controlled hydrodynamic stimuli in a mesopelagic copepod, *Pleuromamma xipias*. *Biol. Bull.* **1999**, *197*, 132–143. [CrossRef] [PubMed]

60. Svetlichny, L.S. Correlation between locomotion parameters and body size at rush swimming in copepods. *J. Gen. Biol.* **1988**, *49*, 401–408. (In Russian)
61. Lenz, P.H.; Hower, A.E.; Hartline, D.K. Force production during pereiopod power strokes in *Calanus finmarchicus*. *J. Mar. Sys.* **2004**, *49*, 133–144. [CrossRef]
62. Buskey, E.J.; Lenz, P.H.; Hartline, D.K. Sensory perception, neurobiology, and behavioral adaptations for predator avoidance in planktonic copepods. *Adapt. Behav.* **2012**, *20*, 57–66. [CrossRef]
63. Bradley, C.J.; Strickler, J.R.; Buskey, E.J.; Lenz, P.H. Swimming and escape behavior in two species of calanoid copepods from nauplius to adult. *J. Plank. Res.* **2013**, *35*, 49–65. [CrossRef]
64. Waggett, R.J. Ecological, Biomechanical and Neurological Correlates of Escape Behavior in Calanoid Copepods. Ph.D. Thesis, University of Texas, Austin, TX, USA, 2005.
65. Catton, K.B.; Webster, D.R.; Yen, J. The effect of fluid viscosity, habitat temperature, and body size on the flow disturbance of *Euchaeta*. *Limnol. Oceanogr. Fluids Environ.* **2012**, *2*, 80–92. [CrossRef]
66. Tanaka, Y. High-speed imaging in copepod behavior. In *Copepods: Diversity, Habitat, and Behavior*; Seuront, L., Ed.; Nova Publishers: Hauppauge, NY, USA, 2014; pp. 145–156.
67. Tuttle, L.J.; Robinson, H.E.; Takagi, D.; Strickler, J.R.; Lenz, P.H.; Hartline, D.K. Going with the flow: Hydrodynamic cues trigger directed escapes from a stalking predator. *J. R. Soc. Interface* **2019**, *16*, 20180776. [CrossRef]
68. Svetlichny, L.S.; Svetlichny, A.S. Measurements of locomotion parameters of copepods fixed to a force element. *Okeanologia* **1986**, *26*, 856–857. (In Russian)
69. Alexander, R.M. The maximum forces exerted by animals. *J. Exp. Biol.* **1985**, *115*, 231–238.
70. Dol'nik, V.R. Allometry of morphology, function, and energy of homoiothermal animal and its physical control. *Zh. Obshch. Biol.* **1982**, *43*, 435–454.
71. Shuleykin, V.V.; Lukyanova, V.S.; Stas, I.I. Comparative dynamics of marine animals. *Proc. USSR Acad. Sci.* **1939**, *22*, 424–429. (In Russian)
72. Svetlichny, L.S.; Stepanov, V.N. On the results of modelling the passive movements in *Calanus helgolandicus*. *Biol. Morya* **1975**, *33*, 61–64.
73. Stepanov, V.N.; Svetlichny, L.S. *Research into the Hydromechanical Characteristics of Planktonic Copepods*; Naukova Dumka: Kiev, Ukraine, 1981; p. 126. (In Russian)
74. Svetlichny, L.S. Hydrodynamic resistance of motionless copepods during their passive sinking in water. *Oceanology* **1983**, *23*, 104–108.
75. Happel, J.; Brenner, H. *Low Reynolds Number Hydrodynamics with Special Applications to Particulate Media*; Prentice-Hall: Englewood Cliffs, NJ, USA, 1965; p. 553.
76. Haury, L.; Weihs, D. Energetically efficient swimming behavior of negatively buoyant zooplankton. *Limnol. Oceanogr.* **1976**, *6*. [CrossRef]
77. Emlet, R.B. Flow fields around ciliated larvae: Effects of natural and artificial tethers. *Mar. Ecol. Prog. Ser.* **1990**, *63*, 211–225. [CrossRef]
78. Schmidt-Nielsen, K. *Scaling: Why is Animal Size So Important?* Cambridge University Press: Cambridge, UK, 1984; p. 241.
79. Jiang, H.; Kiørboe, T. Propulsion efficiency and imposed flow fields of a copepod jump. *J. Exp. Biol.* **2011**, *214*, 476–486. [CrossRef]
80. Murphy, D.W.; Webster, D.R.; Yen, J. A high-speed tomographic PIV system for measuring zooplanktonic flow. *Limnol. Oceanogr. Methods* **2012**, *10*, 1096–1112. [CrossRef]
81. Svetlichny, L.; Hubareva, E. Salinity tolerance of alien copepods *Acartia tonsa* and *Oithona davisae* in the Black Sea. *J. Exp. Mar. Biol. Ecol.* **2014**, *461*, 201–208. [CrossRef]
82. Svetlichny, L.; Hubareva, E.; Isinibilir, M. Comparative trends in respiration rates, sinking and swimming speeds of copepods *Pseudocalanus elongatus* and *Acartia clausi* with comments on the cost of brooding strategy. *J. Exp. Mar. Biol. Ecol.* **2017**, *488*, 24–31. [CrossRef]
83. Svetlichny, L.S.; Hubareva, E.S.; Erkan, F.; Gucu, A.C. Physiological and behavioral aspects of *Calanus euxinus* female (Copepoda, Calanoida) during vertical migration across temperature and oxygen gradients. *Mar. Biol.* **2000**, *137*, 963–971. [CrossRef]
84. Larsen, P.S.; Madsen, C.V.; Riisgård, H.U. Effect of temperature and viscosity on swimming velocity of the copepod *Acartia tonsa*, brine shrimp *Artemia salina* and rotifer *Brachionus plicatilis*. *Aquat. Biol.* **2008**, *4*, 47–54. [CrossRef]

85. Svetlichny, L.; Hubareva, E.; Khanaychenko, A.; Uttieri, M. Salinity and temperature tolerance of the Asian copepod *Pseudodiaptomus marinus* recently introduced into the Black Sea: Sustainability of its invasiveness? *J. Exp. Zool. A Ecol. Integr. Physiol.* **2019**, *331*, 416–426. [PubMed]
86. Marden, J.H. Scaling of maximum net force output by motors used for locomotion. *J. Exp. Biol.* **2005**, *208*, 1653–1664. [CrossRef] [PubMed]
87. Bejian, A.; Marden, J.H. Unifying constructal theory for scale effects in running, swimming and flying. *J Exp. Biol.* **2006**, *209*, 238–248. [CrossRef] [PubMed]
88. Ivlev, V. Energy consumption during the motion of shrimps. *Zool. Zh.* **1963**, *42*, 1465–1471. (In Russian)
89. Torres, J.J.; Childress, J.J. Relationship of oxygen consumption to swimming speed in *Euphausia pacifica*. *Mar. Biol.* **1983**, *74*, 79–86. [CrossRef]
90. Bagshaw, C.R. *Muscle Contraction. Outline Studies in Biology*; Chapman and Hall: London, UK, 1982; p. 127.
91. Gorshkov, V.G. Power and rate of locomotion in animals of different sizes. *J. Gen. Biol.* **1983**, *44*, 661–678. (In Russian)

© 2020 by the authors. Licensee MDPI, Basel, Switzerland. This article is an open access article distributed under the terms and conditions of the Creative Commons Attribution (CC BY) license (http://creativecommons.org/licenses/by/4.0/).

Article

Impacts of Microplastics on the Swimming Behavior of the Copepod *Temora turbinata* (Dana, 1849)

Caroline H. Suwaki, Leandro T. De-La-Cruz and Rubens M. Lopes *

Laboratory of Plankton Systems (LAPS), Oceanographic Institute, University of São Paulo, São Paulo 05508-120, Brazil; caroline.suwaki@usp.br (C.H.S.); leandroticlia@gmail.com (L.T.D.-L.-C.)
* Correspondence: rubens@usp.br; Tel.: +55-11-3091-8963

Received: 4 June 2020; Accepted: 27 June 2020; Published: 30 June 2020

Abstract: Zooplankton are prone to the ingestion of microplastics by mistaking them for prey. However, there is a lack of knowledge about the impacts of microplastic availability on zooplankton behavior. In this study, we investigated the effects of polystyrene microbeads on swimming patterns of the calanoid copepod *Temora turbinata* under laboratory conditions. We acquired high-resolution video sequences using an optical system containing a telecentric lens and a digital camera with an acquisition rate of 20 frames per second. We estimated the mean speed, NGDR (Net-to-Gross Displacement Ratio, a dimensionless single-valued measure of straightness) and turning angle to describe the swimming behavior in three different treatments (control, low and high concentration of microplastics). Our results revealed that swimming speeds decreased up to 40% (instantaneous speed) compared to controls. The NGDR and turning angle distribution of the organisms also changed in the presence of polystyrene microbeads, both at low (100 beads mL^{-1}) and high microplastic concentration (1000 beads mL^{-1}). These results suggest that the swimming behavior of *Temora turbinata* is affected by microbeads.

Keywords: microplastics; zooplankton; swimming behavior; imaging; *Temora turbinata*

---

1. Introduction

Microplastic pollution is now a global concern. Disposal and fragmentation of a wide variety of polymers, followed by their dispersion within large-scale circulation systems, have spread microplastics across the oceans, even to the most remote locations [1–3]. Microplastics (MPs) are particles smaller than 5 mm and can be classified as primary or secondary depending on their origin [4]. Primary MPs include fibers [5], pellets [6] and microspheres from cosmetics and other applications [7]. Secondary MPs are the result of the fragmentation of large plastic objects by a myriad of processes such as UV radiation, mechanical abrasion, and biological degradation by microorganisms [8–10].

Primary and secondary MPs have been recorded in the digestive tract of several marine organisms, including fish [11], annelids [12] and mollusks [13]. Neurological problems [14], hormonal impairment [15], false sensation of satiation, loss of body mass [16] and even death [17] are among the negative impacts of MPs reported to date in marine animals. Because of their size, marine zooplankton have been described as potential MP consumers [8,18]. Copepods are dominant organisms in the marine zooplankton, constituting an average of 80% of the mesozooplankton abundance and representing an efficient link in energy transfer as primary consumers [19]. Copepods participate in essential ecological processes including the biological pump, the microbial food web and nutrient recycling [20,21] and respond quickly to changes in the marine environment such as the introduction of contaminants and man-induced variations in pH, salinity and temperature [22,23]. Although at large spatial scales zooplankton displacement in the water column is strongly affected by currents, their individual swimming behavior is a key factor in achieving high fitness and survival as it controls prey detection, predator avoidance and mating [24].

In the early days of zooplankton feeding studies, MP beads were frequently used as tracers of phytoplankton-sized food to investigate copepod ingestion rates [25] but exposure of copepods to MPs was not treated as a matter of concern in terms of pollution. In this study, we asked the question whether copepod motility would be impaired by the presence of MPs. Changes in the swimming behavior of the calanoid copepod *T. turbinata* were analyzed using high resolution imaging techniques under the influence of varying concentrations of primary MPs.

2. Materials and Methods

2.1. Sampling, Sorting and Exposure of Copepods to MPs

Zooplankton samples were collected in December 2018 from Flamengo inlet, Ubatuba, Brazil (23°31′23″ S, 45°06′13″ W) by means of short oblique tows with a 200-μm mesh-sized plankton net. Animals were transferred to an insulated container and immediately transported to the laboratory, where they were kept in a temperature-controlled room set to match the seawater temperature at the time of sampling (25 °C). Zooplankton organisms were fed for approximately two hours with a saturated concentration of *Isochrysis galbana* under low-intensity aeration. Adult females of the calanoid copepod *Temora turbinata* (Dana, 1849), a dominant species in coastal subtropical waters of the Southwest Atlantic [26], were carefully sorted and 45 healthy individuals were kept in separate 250-mL containers filled with filtered seawater (Whatman® GF/F) for 2 h to acclimatize. Subsequently, 15 females were transferred to each of three 500-mL Nalgene® polycarbonate bottles containing 2×10^4 cells mL^{-1} of *Isochrysis galbana*, set as control (no microspheres) and MP treatments with low (100 mL^{-1}) and high (1000 mL^{-1}) concentrations of surfactant-free polystyrene (PS) latex microspheres (20 μm diameter; Beckman Coulter Inc.). The bottles were sealed with Parafilm M® to avoid bubble formation and transferred to a plankton wheel (1 rpm) for 24 h, at 25 °C. After incubation, 5 individual females belonging to either control or microsphere treatments were retrieved at random from each bottle, inspected for apparent morphological integrity (i.e., only intact and undamaged animals were used), and transferred to a cubic glass vessel containing 500 mL of filtered seawater, where they remained for 15 min to acclimatize before filming with a digital camera system (see below). This time interval was deemed sufficient for complete gut evacuation under the incubation temperature [27]. The sampling and experimental procedures were repeated five times, during four consecutive days.

2.2. Acquisition of Trajectory Data and Analysis of Swimming Behavior

Copepod swimming behavior was recorded in the 5-female batches using a 2D optical system setup consisting of (i) a light-emitting diode (LED) source (660 nm), (ii) a Telecentric lens (0.268X, C-Mount TitanTL, Edmund Optics), and (iii) a 9 MP Basler camera (acA4096-30um with a Sony IMX267 CMOS sensor), operated with an acquisition rate of 20 frames per second (fps). We used a 660 nm LED because most marine organisms, including copepods, are less sensitive to red light, as it is absorbed quickly in the water column [28,29]. Each video observation lasted for 15 min. Image acquisition, spatial coordinate extraction and tracking analysis were performed using software developed by the Laboratory of Planktonic Systems (LAPS/IOUSP). The experimental design is depicted in Figure 1.

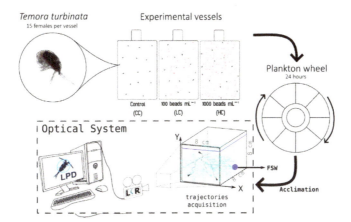

Figure 1. Schematic representation of the experimental setup. CC = Control (with *I. galbana* only); LC = low concentration (*I. galbana* plus 100 beads mL^{-1}); HC = high concentration (*I. galbana* plus 1000 beads mL^{-1}). FSW refers to filtered seawater. LCR (LAPS Camera Recorder) and LPD (LAPS Plankton Detector) are in-house software designed for image acquisition and trajectory extraction, respectively. Image frames were acquired as bidimensional projections (X, Y).

The trajectory extraction process consisted of four main phases: (i) detection of the regions of interest (ROIs) in each frame; (ii) identification of the organisms using a size and contrast range filter; (iii) calculation of the boundary region and centroid position (x, y) of each detected particle in the frame and (iv) generation of 2D trajectories from centroids, having as main guide the closest distance between them in consecutive frames. Additional ROI characteristics, such as area, rotation angle and a time tag, were also obtained.

Trajectories with fewer than 30 frames were discarded to compute behaviorally significant measurements, following Chen and Hwang [30]. We calculated the instantaneous speed, average speed, turning angle and net-to-gross displacement ratio (NGDR). Instantaneous speeds were estimated based on subsequent data points for each trajectory. Average speeds refer to mean speeds for each identified trajectory. The NGDR was calculated from the ratio of the shortest linear distance between start and end points to the total distance traveled. A maximum value of 1 indicates a completely straight path and a value close to 0 indicates a more complex, sinuous path [31]. In this study, we estimated NGDR from segments of fixed size (120 positions) of each track to minimize the effect of spatial scale. Individual turning angles were estimated for every change in movement direction performed by each copepod. Turning angles are reported here as means of individual turning angles after 120 subsequent positions (or image frames) in the bidimensional plane.

Significant differences between controls and treatments were investigated with the Kruskal-Wallis test followed by Dunn's post-hoc test on MATLAB® software R2017a. The significance level was set at $p < 0.05$. Outliers were assigned to data points greater than q3 + w × (q3 − q1) or less than q1 − w × (q3 − q1), where w is the maximum whisker length, and q1 and q3 are the 25th and 75th percentiles of the sample data, respectively.

3. Results

Fifteen videos were acquired containing 883 trajectories, which were visually identified as belonging to three basic types: (i) helical trajectories, when organisms moved forward around an imaginary axis; (ii) rectilinear trajectories, i.e., comparable to a near-straight line; and (iii) curved trajectories, showing intermediate features between the former types (Figure 2). Trajectories with more than 30 points (valid trajectories) represented 74% of the total. No copepod mortality was observed during the experiments.

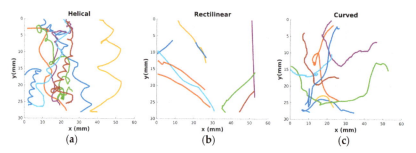

Figure 2. Examples of trajectory types observed in *Temora turbinata* under experimental conditions, including controls and treatments: (**a**) helical-like (n = 158), (**b**) rectilinear (n = 13) and (**c**) curved (n = 482).

The instantaneous swimming speeds (Figure 3) were significantly lower in copepods exposed to plastic microbeads compared to the control, the same trend being observed for mean swimming speeds (Figure 4). The effect of the high MP concentration on the swimming performance of *T. turbinata* was more clearly discerned in the mean speeds (Figure 4; Table 1).

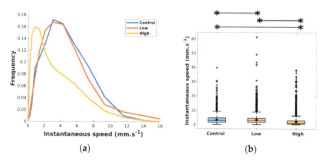

Figure 3. Instantaneous swimming speeds of *T. turbinata* in the control group (with *I. galbana* only) and under MP addition at 100 (Low) and 1000 beads mL^{-1} (High). MP treatments contained *I. galbana* at the same concentration as in the control group. Copepods were observed after being transferred to filtered seawater. (**a**): frequency distribution. (**b**): mean instantaneous speeds, lower and upper limits, and outliers. Horizontal bars on the top of the right panel represent statistical differences between groups.

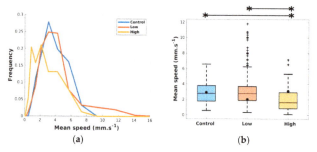

Figure 4. Trajectory-based mean swimming speeds of *T. turbinata* in the control group (with *I. galbana* only) and under MP addition at 100 (Low) and 1000 beads mL^{-1} (High). MP treatments contained *I. galbana* at the same concentration as in the control group. Copepods were observed after being transferred to filtered seawater. (**a**): frequency distribution. (**b**): trajectory mean speeds, lower and upper limits, and outliers. Horizontal bars on the top of the right panel represent statistical differences between groups.

The NGDR also changed under the influence of microbeads, with a decreasing trend of sinuosity in the short-term paths (120 steps) as MP concentration increased (Figure 5). In addition, NGDR segments denoting near-rectilinear steps (i.e., NGDR ~ 1) were more frequent with the increase in microbead concentration.

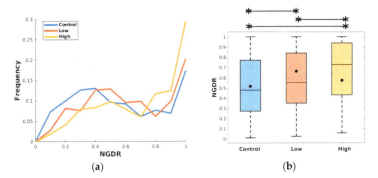

Figure 5. NGDR of *T. turbinata* in the control group (with *I. galbana* only) and under MP addition at 100 (Low) and 1000 beads mL^{-1} (High). MP treatments contained *I. galbana* at the same concentration as in the control group. Copepods were observed after being transferred to filtered seawater. (**a**): frequency distribution. (**b**): mean NGDR, and lower and upper limits. Horizontal bars on the top of the right panel represent statistical differences between groups.

Turning angles displayed by the *T. turbinata* control group had a normal distribution ($\mu = 19.33$, $\sigma = 12.02$), whereas for the LC group a bias existed towards smaller turning angles (Table 1). A more irregular turning angle distribution was noticed for the HC group, which translated into a higher standard deviation compared to the control and the LC treatment (Figure 6; Table 1).

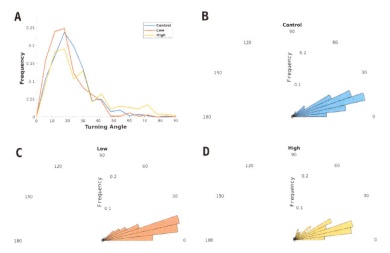

Figure 6. Frequency distribution of the turning angles (in degrees) of *T. turbinata*. (**A**) comparative histogram for the control and treatments, (**B–D**) turning angle frequencies for the control, low and high MP concentrations, respectively. The experimental conditions were the same as indicated in Figures 3–5.

The metrics analyzed here to describe the swimming patterns of T. turbinata in the absence and in the presence of MPs are presented in Table 1 as means and standard deviations. Significant differences are depicted by comparing each treatment to the control.

Table 1. Effects of microplastics in the swimming activity of T. turbinata. CC = Control; LC = Low Concentration (100 beads mL^{-1}) and HC = High Concentration (1000 beads mL^{-1}). Results are presented as means ± standard deviations; significant differences are indicated in bold (Kruskal Wallis test followed by Dunn's post-hoc test, $p < 0.05$).

Treatment	Mean Speed (mm s^{-1})	Mean Instantaneous Speed (mm s^{-1})	NGDR	Turning Angle (°)
CC (n = 187)	2.88 ± 1.35	3.23 ± 2.34	0.52 ± 0.3	19.33 ± 12.05
LC (n = 309)	3.09 ± 1.92	3.19 ± 2.55	**0.57 ± 0.28**	17.69 ± 15.42
HC (n = 157)	**1.99 ± 1.41**	**1.94 ± 2.29**	**0.66 ± 0.28**	**27.22 ± 24.27**

4. Discussion

Swimming is crucial to find prey, encounter mates and avoid predators, affecting zooplankton survival and fitness [32]. However, the impact of MPs on zooplankton swimming behavior is still largely unknown. Here, we found that MPs modify basic patterns of copepod swimming. Our results are consistent with previous findings that MPs cause a decrease in swimming speed of *Artemia* nauplii, *Daphnia* and barnacle larvae at certain concentrations [33–35]. We also observed a reduction in trajectory complexity, as NGDR increased under both MP levels.

Microbeads were often used to mimic phytoplankton cells in grazing experiments and their ingestion by copepods has been recorded [36–40]. Polystyrene spheres are not necessarily ingested by copepods when offered alone, while phytoplankton cells are consumed at rates up to 3 times higher than non-living items when both are available, denoting a clear particle selectivity pattern [41]. However, algal ingestion by copepods decreases when high MP concentrations are offered in combination with natural food [42]. For instance, Ayukai [43] reported that phytoplankton ingestion by the copepod *Acartia clausi* decreased in the presence of MPs of similar size (15.7 μm) and concentration (1140 beads mL^{-1}) used in the present study. The food offered to T. turbinata in our experiments (*I. galbana*) had a size spectrum (5–7 μm) within the capture range previously reported for this copepod genus [44] and algal concentration was kept constant in the different MP treatments and control. Thus, it is unlikely that behavioral differences observed in our data resulted from varying levels of "satiation" associated with MP consumption at different concentrations. This is reinforced by the fact that copepod trajectories were recorded for control and MP treatments while animals were swimming in filtered seawater, after full gut evacuation, minimizing potential bias between starved and fed individuals.

MP ingestion has been shown to cause enzyme (cholinesterase and catalase) impairment in microcrustaceans, affecting the cholinergic system and causing oxidative stress [34]. Energy deficit, reduced growth, and low fertility rates have also been reported as negative impacts of MP consumption [45,46]. In addition, MPs may accumulate on the outer surface of copepods and become entangled around the antennae, swimming legs, and feeding appendages [42], probably affecting swimming performance and other functions. Although no observation of either metabolic damage or particle trapping on copepod surfaces is available from our experiments, the reduction in swimming speed and the high NGDRs and turning angles observed after MP exposure may be interpreted as evidence of a direct impact from plastic microbeads. Also, under "normal" experimental conditions (i.e., without MP exposure) T. turbinata increases its average swimming speed in the presence of food [47], but we found an opposite trend for copepods exposed to microbeads.

The MP concentrations (100 and 1000 beads mL^{-1}) applied in our experiments were similar or lower than those utilized in previous investigations of MP consumption by copepods [42,43,45]. The extent to which such concentrations are realistic for the natural environment, even under a MP accumulation scenario in the oceans, is a matter of debate because so far most marine studies have targeted MP particles larger than 300 µm [42], although evidence exists that fibrous polymers in the range of ~20 µm are present in coastal sediments [6]. Sampling with an 80-µm mesh-sized plankton net yielded a peak concentration of a mere 0.1 MPs mL^{-1} in the west coast of Sweden [48], but such estimate is probably related to particles larger than 100 µm or more, because of mesh size selectivity [49]. Considering the sampling constraints and the fact that plankton abundance in the size range of 80–100 µm is typically three to four orders of magnitude lower than in the 15–20 µm size range [50], a concentration of 100–1000 mL^{-1} for 20-µm sized spherical MPs would not be surprising for a coastal environment, particularly near densely populated urban areas. Interestingly, a recent study found that MPs smaller than 25 µm accounted for about 70% of the total number of airborne, plastic-derived particles settled in pristine continental areas of the United States [51]. Thus, in case atmospheric transport is shown to represent a relevant MP input to the oceans, it is likely that small-sized particles will account for a major proportion of the total plastic load in surface waters.

As they move up and down the food web, MPs potentially affect diverse ecosystem processes [52,53], both locally and remotely. For instance, Cole and Galloway [45] observed that fecal pellets of the copepod *Calanus helgolandicus* exposed to 20 µm MPs (1000 beads mL^{-1}) had a 2.25-fold reduction in their sinking rate, which translated to an increase of 53 days for pellets to reach the seafloor, considering the average depth of the ocean [54]. In addition, the relative energy cost of locomotion increases as the size of the organism decreases [55], meaning that for small zooplankton such as copepods, changes in swimming behavior due to MP influence likely affect vertical migration and prey-predator interactions, modifying the amount of energy available for the different trophic levels [38,56]. Such potential impact of MPs on copepods may thus cause large-scale alterations in the carbon flow in marine environments. Therefore, it is imperative that the impacts of MP on copepod behavior are elucidated and considered in trophic models and biogeochemical studies in the future.

5. Conclusions

This study shows that the availability of polystyrene microspheres modified the swimming performance of the pelagic copepod *T. turbinata* at both low and high MP concentrations. Changes in copepod swimming behavior as a response to the presence of microplastics may elicit individual-based effects leading to potential impacts on several ecological traits, including bottom-up transference of MP particles to higher trophic levels.

Author Contributions: Conceptualization, C.H.S. and R.M.L.; laboratory analysis, C.H.S.; software and data analysis, L.T.D.-L.-C.; data interpretation, C.H.S., L.T.D.-L.-C. and R.M.L.; writing—initial draft preparation, C.H.S. and L.T.D.-L.-C.; writing—final draft preparation, review, and editing, R.M.L.; supervision, R.M.L. All authors have read and agreed to the published version of the manuscript.

Funding: This research received no external funding. R.M.L. is a CNPq Researcher.

Acknowledgments: The authors gratefully acknowledge the operational support by the staff of the University of Sao Paulo's marine research station "Base Clarimundo de Jesus" and the colleagues Yonara Garcia and Felipe Neves for discussions during early stages of the study. Two anonymous reviewers provided invaluable contributions with their comments and suggestions.

Conflicts of Interest: The authors declare no conflict of interest.

References

1. Barnes, D.K.; Galgani, F.; Thompson, R.C.; Barlaz, M. Accumulation and fragmentation of plastic debris in global environments. *Philos. Trans. R. Soc. B Biol. Sci.* **2009**, *364*, 1985–1998. [CrossRef]
2. Eriksen, M.; Lebreton, L.C.; Carson, H.S.; Thiel, M.; Moore, C.J.; Borerro, J.C.; Galgani, F.; Ryan, P.G.; Reisser, J. Plastic pollution in the world's oceans: More than 5 trillion plastic pieces weighing over 250,000 tons afloat at sea. *PLoS ONE* **2014**, *9*, e111913. [CrossRef]
3. Zarfl, C.; Matthies, M. Are marine plastic particles transport vectors for organic pollutants to the Arctic? *Mar. Pollut. Bull.* **2010**, *60*, 1810–1814. [CrossRef]

4. Betts, K. *Why Small Plastic Particles May Pose a Big Problem in the Oceans*; ACS Publications: Washington, DC, USA, 2008.
5. Napper, I.E.; Thompson, R.C. Release of synthetic microplastic plastic fibres from domestic washing machines: Effects of fabric type and washing conditions. *Mar. Pollut. Bull.* **2016**, *112*, 39–45. [CrossRef]
6. Andrady, A.L. Microplastics in the marine environment. *Mar. Pollut. Bull.* **2011**, *62*, 1596–1605. [CrossRef]
7. Fendall, L.S.; Sewell, M.A. Contributing to marine pollution by washing your face: Microplastics in facial cleansers. *Mar. Pollut. Bull.* **2009**, *58*, 1225–1228. [CrossRef]
8. Cole, M.; Lindeque, P.; Halsband, C.; Galloway, T.S. Microplastics as contaminants in the marine environment: A review. *Mar. Pollut. Bull.* **2011**, *62*, 2588–2597. [CrossRef]
9. Song, Y.K.; Hong, S.H.; Jang, M.; Han, G.M.; Jung, S.W.; Shim, W.J. Combined effects of UV exposure duration and mechanical abrasion on microplastic fragmentation by polymer type. *Environ. Sci. Technol.* **2017**, *51*, 4368–4376. [CrossRef]
10. Ter Halle, A.; Ladirat, L.; Gendre, X.; Goudounèche, D.; Pusineri, C.; Routaboul, C.; Tenailleau, C.; Duployer, B.; Perez, E. Understanding the fragmentation pattern of marine plastic debris. *Environ. Sci. Technol.* **2016**, *50*, 5668–5675. [CrossRef]
11. Lusher, A.; Mchugh, M.; Thompson, R. Occurrence of microplastics in the gastrointestinal tract of pelagic and demersal fish from the English Channel. *Mar. Pollut. Bull.* **2013**, *67*, 94–99. [CrossRef]
12. Browne, M.A.; Niven, S.J.; Galloway, T.S.; Rowland, S.J.; Thompson, R.C. Microplastic moves pollutants and additives to worms, reducing functions linked to health and biodiversity. *Curr. Biol.* **2013**, *23*, 2388–2392. [CrossRef] [PubMed]
13. Van Cauwenberghe, L.; Janssen, C.R. Microplastics in bivalves cultured for human consumption. *Environ. Pollut.* **2014**, *193*, 65–70. [CrossRef]
14. Barboza, L.G.A.; Vieira, L.R.; Branco, V.; Figueiredo, N.; Carvalho, F.; Carvalho, C.; Guilhermino, L. Microplastics cause neurotoxicity, oxidative damage and energy-related changes and interact with the bioaccumulation of mercury in the European seabass, Dicentrarchus labrax (Linnaeus, 1758). *Aquat. Toxicol.* **2018**, *195*, 49–57. [CrossRef] [PubMed]
15. Facciolo, R.M.; Madeo, M.; Giusi, G.; Carelli, A.; Canonaco, M. Effects of the xenoestrogen bisphenol A in diencephalic regions of the teleost fish Coris julis occur preferentially via distinct somatostatin receptor subtypes. *Brain Res. Bull.* **2005**, *65*, 267–273.
16. Welden, N.A.; Cowie, P.R. Environment and gut morphology influence microplastic retention in langoustine, Nephrops norvegicus. *Environ. Pollut.* **2016**, *214*, 859–865. [CrossRef]
17. Derraik, J.G. The pollution of the marine environment by plastic debris: A review. *Mar. Pollut. Bull.* **2002**, *44*, 842–852. [CrossRef]
18. Cole, M.; Galloway, T.S. Ingestion of nanoplastics and microplastics by Pacific oyster larvae. *Environ. Sci. Technol.* **2015**, *49*, 14625–14632. [CrossRef]
19. Turner, J.T. The importance of small planktonic copepods and their roles in pelagic marine food webs. *Zool. Stud.* **2004**, *43*, 255–266.
20. Turner, J.T. Zooplankton fecal pellets, marine snow, phytodetritus and the ocean's biological pump. *Prog. Oceanogr.* **2015**, *130*, 205–248. [CrossRef]
21. Zöllner, E.; Hoppe, H.-G.; Sommer, U.; Jürgens, K. Effect of zooplankton-mediated trophic cascades on marine microbial food web components (bacteria, nanoflagellates, ciliates). *Limnol. Oceanogr.* **2009**, *54*, 262–275. [CrossRef]
22. Gannon, J.E.; Stemberger, R.S. Zooplankton (especially crustaceans and rotifers) as indicators of water quality. *Trans. Am. Microsc. Soc.* **1978**, *97*, 16–35. [CrossRef]
23. Zannatul, F.; Muktadir, A. A review: Potentiality of zooplankton as bioindicator. *Am. J. Appl. Sci.* **2009**, *6*, 1815–1819.
24. Kiørboe, T. *A Mechanistic Approach to Plankton Ecology*; Princeton University Press: Princeton, NJ, USA, 2008.
25. Frost, B.W. Feeding behavior of Calanus pacificus in mixtures of food particles 1. *Limnol. Oceanogr.* **1977**, *22*, 472–491. [CrossRef]
26. Lopes, R.M.; Brandini, F.P.; Gaeta, S.A. Distribution patterns of epipelagic copepods off Rio de Janeiro (SE Brazil) in summer 1991/1992 and winter 1992. *Hydrobiologia* **1999**, *411*, 161–174. [CrossRef]
27. Irigoien, X. Gut clearance rate constant, temperature and initial gut contents: A review. *J. Plankton Res.* **1998**, *20*, 997–1003. [CrossRef]

28. Buskey, E.J.; Baker, K.S.; Smith, R.C.; Swift, E. Photosensitivity of the oceanic copepods Pleuromamma gracilis and Pleuromamma xiphias and its relationship to light penetration and daytime depth distribution. *Mar. Ecol. Prog. Ser.* **1989**, *55*, 207–216. [CrossRef]
29. Nayak, A.R.; McFarland, M.N.; Sullivan, J.M.; Twardowski, M.S. Evidence for ubiquitous preferential particle orientation in representative oceanic shear flows. *Limnol. Oceanogr.* **2018**, *63*, 122–143. [CrossRef]
30. Chen, M.-R.; Hwang, J.-S. The swimming behavior of the calanoid copepod Calanus sinicus under different food concentrations. *Zool. Stud.* **2018**, *57*, e13.
31. Buskey, E.J. Swimming pattern as an indicator of the roles of copepod sensory systems in the recognition of food. *Mar. Biol.* **1984**, *79*, 165–175. [CrossRef]
32. Visser, A.W.; Kiørboe, T. Plankton motility patterns and encounter rates. *Oecologia* **2006**, *148*, 538–546. [CrossRef]
33. De Felice, B.; Sabatini, V.; Antenucci, S.; Gattoni, G.; Santo, N.; Bacchetta, R.; Ortenzi, M.A.; Parolini, M. Polystyrene microplastics ingestion induced behavioral effects to the cladoceran Daphnia magna. *Chemosphere* **2019**, *231*, 423–431. [CrossRef] [PubMed]
34. Gambardella, C.; Morgana, S.; Ferrando, S.; Bramini, M.; Piazza, V.; Costa, E.; Garaventa, F.; Faimali, M. Effects of polystyrene microbeads in marine planktonic crustaceans. *Ecotoxicol. Environ. Saf.* **2017**, *145*, 250–257. [CrossRef] [PubMed]
35. Wang, X.; Liu, L.; Zheng, H.; Wang, M.; Fu, Y.; Luo, X.; Li, F.; Wang, Z. Polystyrene microplastics impaired the feeding and swimming behavior of mysid shrimp Neomysis japonica. *Mar. Pollut. Bull.* **2020**, *150*, 110660. [CrossRef] [PubMed]
36. Desforges, J.-P.W.; Galbraith, M.; Ross, P.S. Ingestion of microplastics by zooplankton in the Northeast Pacific Ocean. *Arch. Environ. Contam. Toxicol.* **2015**, *69*, 320–330. [CrossRef]
37. Powell, M.D.; Berry, A. Ingestion and regurgitation of living and inert materials by the estuarine copepod Eurytemora affinis (Poppe) and the influence of salinity. *Estuar. Coast. Shelf Sci.* **1990**, *31*, 763–773. [CrossRef]
38. Setälä, O.; Fleming-Lehtinen, V.; Lehtiniemi, M. Ingestion and transfer of microplastics in the planktonic food web. *Environ. Pollut.* **2014**, *185*, 77–83. [CrossRef]
39. Van Alstyne, K.L. Effects of phytoplankton taste and smell on feeding behavior of the copepod Centropages hamatus. *Mar. Ecol. Prog. Ser.* **1986**, *34*, 187–190. [CrossRef]
40. Wright, S.L.; Thompson, R.C.; Galloway, T.S. The physical impacts of microplastics on marine organisms: A review. *Environ. Pollut.* **2013**, *178*, 483–492. [CrossRef]
41. Paffenhöfer, G.-A.; Van Sant, K.B. The feeding response of a marine planktonic copepod to quantity and quality of particles. *Mar. Ecol. Prog. Ser.* **1985**, *27*, 55–65. [CrossRef]
42. Cole, M.; Lindeque, P.; Fileman, E.; Halsband, C.; Goodhead, R.; Moger, J.; Galloway, T.S. Microplastic ingestion by zooplankton. *Environ. Sci. Technol.* **2013**, *47*, 6646–6655. [CrossRef]
43. Ayukai, T. Discriminate feeding of the calanoid copepod Acartia clausi in mixtures of phytoplankton and inert particles. *Mar. Biol.* **1987**, *94*, 579–587. [CrossRef]
44. Kleppel, G.; Burkart, C.; Carter, K.; Tomas, C. Diets of calanoid copepods on the West Florida continental shelf: Relationships between food concentration, food composition and feeding activity. *Mar. Biol.* **1996**, *127*, 209–217. [CrossRef]
45. Cole, M.; Lindeque, P.; Fileman, E.; Halsband, C.; Galloway, T.S. The impact of polystyrene microplastics on feeding, function and fecundity in the marine copepod Calanus helgolandicus. *Environ. Sci. Technol.* **2015**, *49*, 1130–1137. [CrossRef]
46. Ogonowski, M.; Schür, C.; Jarsén, Å.; Gorokhova, E. The effects of natural and anthropogenic microparticles on individual fitness in Daphnia magna. *PLoS ONE* **2016**, *11*, e0155063. [CrossRef] [PubMed]
47. Wu, C.-H.; Dahms, H.-U.; Buskey, E.J.; Strickler, J.R.; Hwang, J.-S. Behavioral interactions of the copepod Temora turbinata with potential ciliate prey. *Zool. Stud.* **2010**, *49*, 157–168.
48. Lozano, R.; Mouat, J. Marine litter in the North-East Atlantic Region: Assessment and Priorities for Response. Available online: https://www.semanticscholar.org/paper/Marine-litter-in-the-Northeast-Atlantic-Region%3A-and-Other-Lozano/411a7c97d29b5874d6799c58f7983149bcb049c6 (accessed on 1 June 2020).
49. Nichols, J.; Thompson, A. Mesh selection of copepodite and nauplius stages of four calanoid copepod species. *J. Plankton Res.* **1991**, *13*, 661–671. [CrossRef]

50. Lombard, F.; Boss, E.; Waite, A.M.; Vogt, M.; Uitz, J.; Stemmann, L.; Sosik, H.M.; Schulz, J.; Romagnan, J.-B.; Picheral, M. Globally consistent quantitative observations of planktonic ecosystems. *Front. Mar. Sci.* **2019**, *6*, 196. [CrossRef]
51. Brahney, J.; Hallerud, M.; Heim, E.; Hahnenberger, M.; Sukumaran, S. Plastic rain in protected areas of the United States. *Science* **2020**, *368*, 1257–1260. [CrossRef]
52. Carbery, M.; O'Connor, W.; Palanisami, T. Trophic transfer of microplastics and mixed contaminants in the marine food web and implications for human health. *Environ. Int.* **2018**, *115*, 400–409. [CrossRef]
53. Wang, W.; Gao, H.; Jin, S.; Li, R.; Na, G. The ecotoxicological effects of microplastics on aquatic food web, from primary producer to human: A review. *Ecotoxicol. Environ. Saf.* **2019**, *173*, 110–117. [CrossRef]
54. Galloway, T.S.; Cole, M.; Lewis, C. Interactions of microplastic debris throughout the marine ecosystem. *Nat. Ecol. Evol.* **2017**, *1*, 1–8. [CrossRef] [PubMed]
55. Hansen, A.N.; Visser, A.W. Carbon export by vertically migrating zooplankton: An optimal behavior model. *Limnol. Oceanogr.* **2016**, *61*, 701–710. [CrossRef]
56. Cedervall, T.; Hansson, L.-A.; Lard, M.; Frohm, B.; Linse, S. Food chain transport of nanoparticles affects behaviour and fat metabolism in fish. *PLoS ONE* **2012**, *7*, e32254. [CrossRef] [PubMed]

© 2020 by the authors. Licensee MDPI, Basel, Switzerland. This article is an open access article distributed under the terms and conditions of the Creative Commons Attribution (CC BY) license (http://creativecommons.org/licenses/by/4.0/).

Article

Chemical Signaling in the Turbulent Ocean—Hide and Seek at the Kolmogorov Scale

Erik Selander [1,*], **Sam T. Fredriksson** [2] **and Lars Arneborg** [2,*]

[1] Department of Marine Sciences, University of Gothenburg, Carl Skottsbergs g 22B, SE 413 19 Gothenburg, Sweden
[2] Department of Research and Development, Swedish Meteorological and Hydrological Institute, Sven Källfelts Gata 15, SE 426 71 Västra Frölunda, Sweden; sam.fredriksson@smhi.se
* Correspondence: erik.selander@marine.gu.se (E.S.); lars.arneborg@smhi.se (L.A.)

Received: 4 March 2020; Accepted: 13 April 2020; Published: 21 April 2020

Abstract: Chemical cues and signals mediate resource acquisition, mate finding, and the assessment of predation risk in marine plankton. Here, we use the chemical properties of the first identified chemical cues from zooplankton together with in situ measurements of turbulent dissipation rates to calculate the effect of turbulence on the distribution of cues behind swimmers as well as steady state background concentrations in surrounding water. We further show that common zooplankton (copepods) appears to optimize mate finding by aggregating at the surface in calm conditions when turbulence do not prevent trail following. This near surface environment is characterized by anisotropic turbulence and we show, using direct numerical simulations, that chemical cues distribute more in the horizontal plane than vertically in these conditions. Zooplankton may consequently benefit from adopting specific search strategies near the surface as well as in strong stratification where similar flow fields develop. Steady state concentrations, where exudation is balanced by degradation develops in a time scale of ~5 h. We conclude that the trails behind millimeter-sized copepods can be detected in naturally occurring turbulence below the wind mixed surface layer or in the absence of strong wind. The trails, however, shorten dramatically at high turbulent dissipation rates, above ~10^{-3} cm^2 s^{-3} (10^{-7} W kg^{-1}).

Keywords: Kolmogorov; turbulence; copepod; chemosensory; signaling; zooplankton

1. Introduction

The open ocean is a dilute environment. Organisms have to process large volumes of water to acquire resources and find mates [1]. At the same time, predation rates are high, and organisms have to trade resource acquisition against predator avoidance. These contradictory needs drive evolution of advanced sensory systems to improve detection of both resources and threats. The vast majority of marine plankton depend on hydro-mechanical and chemosensory information. Both depend on the fluid flow regime. Here, we focus on the effects of turbulence on chemical sensing and signaling at the length scale characteristic for the transition between viscous and turbulent flows, the Kolmogorov length scale.

The most numerous organisms in the sea, bacteria, experience a fully viscous flow regime. The transport of solutes is mainly governed by diffusion, and the effect of turbulence often marginal. Bacteria use simple behavioral algorithms to maneuver the chemical landscape in a way that is fairly well understood [2]. In principle, the rate of change of direction is altered in response to concentration gradients in a way that allows bacteria to navigate towards, or away from sources. Sensing and signaling in slightly larger organisms is, however, more complex. The most abundant multicellular animals in the sea, copepods, are all in this size range, 100 µm to a few mm, and may even traverse the Kolmogorov scale when growing up from larval stages to adults (Figure 1).

Figure 1. Artistic interpretation of the distribution of solutes around small (μm, e.g., bacteria and small phytoplankton) and intermediate sized particles (mm) such as motile copepods in the transition from viscous to turbulent regimes (from left to right). At micrometer scale the solute distribution is largely driven by diffusion. The chemical wake behind motile cells is eroded by diffusion faster than it forms and solute distribution is only moderately affected by turbulence. At slightly larger scales, motile organisms leave chemical trails in their wake that can be detected tens to hundreds body lengths away [3]. At higher turbulent dissipation rates however, the trails will be deformed and thinned by turbulence in a way that reduce detectability [4]. Local gradients will also be more rapidly mixed into background concentrations at high turbulence (Illustration: Jan Heuschele).

While it has long been known that copepods emit and receive chemical cues involved in mate finding [3,5], resource acquisition, and predator avoidance [6], the cueing compounds have remained largely unknown. As a consequence, fundamental parameters such as exudation rate, degradation, diffusivity, and sensitivity thresholds have been lacking. This is one of the reasons why the literature holds contrasting views on whether zooplankton can exploit chemical trails to increase encounter rates in nature, or if turbulence erodes the trace too fast (see e.g., [3,7]) in the ocean. Recently, the first signaling compounds from copepods were identified as a group of polar lipids, copepodamides [8]. Copepodamides induce defensive traits such as toxin production, bioluminescence, or colony size plasticity in prey organisms [8–10]. The exudation rate, degradation rate, and sensitivity threshold have now been established for copepodamides [8,9].

The purpose of the present study is to combine the empirical data on physical properties of cueing compounds with in-situ measurements of turbulent dissipation rates to test the theories on the effect of turbulence on sensing and signaling in the open ocean. Moreover, copepods form near surface reproductive swarms [11]. Near the surface, however, the turbulence is typically anisotropic [12–15]. We explore the effect of anisotropic turbulence on chemosensing near the surface using direct numerical simulations.

Theoretical Background

The physics of trail formation behind zooplankton has been thoroughly addressed by others. A simple model for the solute concentration behind a copepod in calm water is that of a point source moving at constant velocity through a diffusive medium [3]. The diffusion problem can be solved analytically [16] and the solution can be expressed as

$$C = \frac{Q}{4\pi Dz} exp\left(-\frac{Ur^2}{4Dz}\right) \qquad (1)$$

where z is the along-track distance behind the copepod and r is the radial distance from the centerline of the trail, remaining parameters defined in Table 1. The same expression can be used for chemical trails behind particles falling through the water column with velocity U [17]. An organism with detection limit C^* can detect the trail at distance

$$z_0 = \frac{Q}{4\pi D C^*} \quad (2)$$

It is noteworthy that the detection length is independent of velocity. For a fast copepod, the trail is thinner and diffuses faster than for a slow, but the detectable trail length is the same [1].

In a turbulent environment, the trail will be stretched and thinned by turbulent straining. At scales smaller than the Kolmogorov scale, the viscous straining dominates the relative displacement of particles, and since this is the movement that removes most of the kinetic energy from the flow, there is a direct relationship between straining, viscosity, and dissipation of turbulent kinetic energy. The rate of straining, γ, scales as:

$$\gamma = \left(\frac{\varepsilon}{6\nu}\right)^{1/2} \quad (3)$$

Diffusivity tends to widen the trail whereas advective thinning due to straining tends to narrow it. A balance between diffusive widening and thinning due to strain is obtained when the trail width is about equal to the Batchelor scale, $B = (\nu D^2/\varepsilon)^{\frac{1}{4}}$ [18]. Note that this is the scale of the actual trail width, which is different from the detectable width that decays due to decreasing trail centerline concentrations. At this width, the rate of change of trail concentration scales as

$$\frac{dC}{dt} \propto -D\frac{C}{B^2} \propto -\gamma C, \quad (4)$$

i.e., the inverse of γ is a measure for the decay time-scale of the trail concentration in a turbulent flow. Another time scale of importance for the trail is the time it takes before a trail becomes undetectable in calm water,

$$T_0 = \frac{z_0}{U}. \quad (5)$$

Table 1. Parameters and values used in calculations.

Parameter	Symbol	Value	Reference
Copepod length	L	0.1 cm	
Copepod swimming speed	U	0.05–1 cm s^{-1}	[19]
Copepod concentration	n	10–12 ind l^{-1}	sharkweb.smhi.se
Emission rate of copepodamides *	Q	17 pmol ind^{-1} d^{-1}	[8]
Sensitivity threshold for copepods	C*	1 nM	[6] and references there in
Sensitivity threshold for microplankton	C^p	20–80 fM	[9]
Degradation rate	k	0.21h^{-1}(C$_t$ = C$_0$e$^{-0.21t}$)	[20]
Molecular diffusivity	D	3 × 10^{-6} cm^2 s^{-1}	[21]
Kinematic viscosity	ν	10^{-2} cm^2 s^{-1}	
Dissipation rate of turbulent kinetic energy	ε	10^{-6}–10^{3} cm^2 s^{-3}	[22]

* Calculated from weight specific copepodamide production [8] to the copepod species (*Centropages typicus*) used in [3].

The main, non-dimensional, parameter that determines the influence of turbulence on a trail is the relation between these two time scales,

$$VJ = \gamma T_0 \quad (6)$$

which we call the *VJ* number after Visser and Jackson [4]. Visser and Jackson [4] used a statistical model of isotropic turbulence and a relatively simple trail model to show that the trail characteristics are little influenced by turbulence for *VJ* < 1 whereas the trail area, volume, and decay time scale is much smaller than in calm water for *VJ* > 100. It is noteworthy that the actual length of the trail remains as long in turbulent as in calm water, but it curls up, becomes thinner, and decays much quicker and closer to the copepod in turbulent water. In other words, the material lines are elongated in turbulent water, but the elongation causes a thinning that increases the diffusive dilution so much that the material trail length remains the same. Visser and Jacksson [4] presented the following equation for the detection time scale of a trail in turbulence

$$T = T_0 \frac{\ln(VJ+1)}{VJ} \tag{7}$$

which approaches T_0 for small VJ and 0 for large VJ. A similar relationship can be obtained between the trail length in calm water, z_0, and the detection distance behind the copepod. However, these results still remain to be validated in laboratory or in direct numerical simulations. It is also limited to isotropic turbulence and do therefore do not necessarily apply near the surface or in strongly stratified fluid.

2. Materials and Methods

2.1. Calculations of Trail Lengths and Background Concentrations

The empirical data on production, degradation, and sensitivity thresholds were collected from the literature (Table 1). Observations and samples of surface swarms of copepods were obtained by bucket-sampling the densest part of two swarms observed on the Swedish west coast (58°14′59.5″N 11°26′43.8″E and 58°52′33.5″N 11°08′43.2″E). The copepods were preserved in Lugols's solution. All copepods in 65 mL (*Centropages*) or three times 50 mL (*Calanus*) were counted and prosome length (*Calanus*) or sex (*Centropages*) was determined under a dissection microscope.

Degradation rates of copeopdamides in water is from unpublished work by Arias et al. (2020) who measured degradation rates at 19 °C in seawater.

2.2. Direct Numerical Simulations

Direct numerical simulations (DNS) are here used to study the turbulent flow at the sea-air interface and how the turbulence affects an inert tracer such as a chemical signal. The turbulent flow is here driven by wind stress and thermal convection. The surface shear stress condition is set to give the surface-shear based Reynolds number $Re_*^* = u_* H/\nu = 120$, where $u_* = \sqrt{\tau_0/\rho}$ is the friction velocity, H is the length scale here given by the domain depth, and ν is the kinematic viscosity. τ_0 is the surface shear stress and ρ is the density. The water-side friction velocity can be converted to wind speed U_{10} at 10 m height using first $u_{*,a} = u_* \sqrt{\rho/\rho_a}$ and then

$$\frac{U(z)}{u_{*,a}} = \kappa^{-1} \ln\left(\frac{u_{*,a} z}{\nu_a}\right) + 5.7 \tag{8}$$

valid for neutral conditions in the atmosphere [23]. This gives $U_{10} \approx 1.3$ m s^{-1} which together with the used natural convection $Q_0 = 100$ W m^{-2} give conditions similar to a summer evening with clear skies and low wind conditions, comparable to the conditions the two near surface swarming events were observed in.

The modeled 0.1 × 0.4 × 1 m volume is shown in Figure 2. The stream direction and depth are denoted x and z. y is perpendicular to stream direction. The domain size is $3\pi H \times \pi H \times H$ in the x-, y- and z-directions where $H = 0.1204$ m. It is discretized using 1206 × 402 × 96 cells. The cells are equidistant in the x- and y-directions (~0.94 mm). The mesh distribution is stretched in the z-direction, where the smallest and largest cell heights, closest to the surface and bottom respectively, are ~0.098 mm and ~1.96 mm.

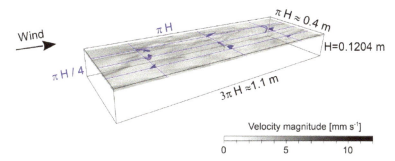

Figure 2. The modeled volume showing iso-concentrations of signal concentration (blue) around nine near surface point sources and the magnitude of flow velocity at the depth of the point source (0.995 cm). This plane is made transparent in order to visualize which part of the signal plumes are above and below its point of dispersal. The cross- and stream-wise positions of the point sources are given as the interceptions of the blue lines. The distances are $\pi H/4$ and πH, respectively.

The incompressible Navier–Stokes equations are solved using the standard Boussinesq approximation together with a thermal energy equation [12]. The transport equation for the passive signal is used to calculate the signal field.

$$\frac{\partial S}{\partial t} + \mathbf{U}\cdot\nabla S = D\nabla^2 S + \varnothing_S \tag{9}$$

where S is the signal concentration, t is time, and $\mathbf{U}$ is the fluid velocity. D is the signal molecular diffusivity and $\varnothing_S$ is a spatially and temporally constant source terms added in nine positions at the same depth of 0.995 cm. The depth represents typical depth for copepod swarms and coincides with stream velocity $\bar{u} \approx 1$ mm s^{-1} similar to copepod swimming speed. It is also interesting to see how anisotropic flow close to the surface influences the dispersal of the signal. The point sources are distributed evenly in both cross- and streamwise directions. This results in a $\Delta y = H\cdot\pi/4 = 0.094\ m$ and $\Delta x = H\cdot\pi = 0.378\ m$. These distances are chosen to be of the same order as the length scale H and in the streamwise direction to be long enough to enable a reasonable sampling time before the different sampling sources start to interfere with each other. The diffusivity is chosen to be the same as thermal diffusivity since the small-scale gradients induced by the much smaller diffusivities of copepodamides cannot be resolved with the mesh resolution used in these simulations.

The surface boundary is assumed to be flat assuming that the surface deflection is negligible. The signal boundary condition for the surface and bottom is $\partial S/\partial z = 0$ which represent no signal exchange through these boundaries. Finally, the flow is subject to periodic (cyclic) boundary conditions in the horizontal (x- and y-) directions.

The simulations are carried out using a collocated finite volume approach with OpenFOAM, which is an open-source computational fluid dynamics tool. The computational time step Δt is set dynamically ensuring that the Courant–Friedrich–Lewy number, $CFL = \Delta t|U|/\Delta l < 0.5$, which results in $\Delta t \approx 0.025\ s$. Here $|U|$ is the magnitude of the velocity in a cell and Δl is the length of the cell in the velocity direction. The mesh resolution and time discretization are further discussed by [13].

Sampling of the signal concentrations is done under 60 s after 40 s of continuous seeding. The mean concentration is calculated by first superimposing an area around each of the nine seeding positions (Figure 2). The mean concentration is deducted, and a temporal average performed (Appendix A). The sampling period is short to avoid that the signal from one source interfere with adjacent sources. The values should consequently be considered qualitative rather than quantitative.

3. Results and Discussion

3.1. Mate Finding is Easier in 2 Dimensions than in 3

The most obvious way to increase encounter rate is to gather at the surface (or the bottom). This reduce the problem to two rather than three dimensions and at the same time increase concentration. Dense swarms of copepods are also observed on quiet summer days in the study area (Swedish west coast), and in e.g., freshwater copepods [11]. Among the species we observed to form swarms were *Centropages typicus*, *Acartia sp*, *Oithona sp*, *Pseudocalanus sp*, and *Calanus sp*. The swarms are usually dominated by adults of a single species (Figure 3) suggesting a reproductive role. A 65 mL sample from a *Centropages* swarm for example contained 343 males and 39 females out of which approximately half (17) had a spermatophore attached and where thus mated. Three 50 mL samples from the center of a *Calanus* swarm contained on average 1468 copepods each, corresponding to >29,000 copepods L^{-1}, also dominated by the large life stages although these were not sexed (Figure 3).

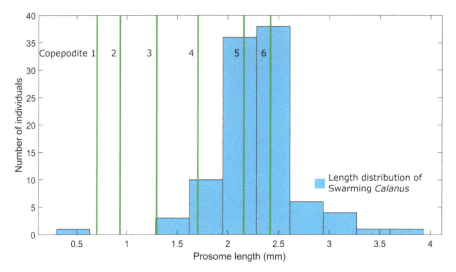

Figure 3. Length distribution of swarming *Calanus sp*. The green lines represent the average size of the copepodite stages (1–5 are, juveniles 6 is sexually mature adults) from monitoring data (www.sharkweb.smhi.se). The swarms are dominated by adults and late stage copepodites supporting the reproductive role of swarms. This particular swarm reached extreme densities >29,000 copepods L^{-1}. Most swarms are less dense.

Swarming is a powerful strategy to facilitate mate finding provided that there is a reliable cue to synchronize the ascent to the surface. Still, however, chemical information is useful for close range search and selective mating.

3.2. Effect of Turbulence on Chemical Trails

When applying the empirical data in Table 1 on Equation (2), copepods can easily produce trails trackable over 30–50 body-lengths in calm water, or >100 body lengths for larger copepods. Turbulence, however, deforms and dilute the chemical trails, depending on the *VJ* number as described in the introduction.

From Equations (1)–(6), one can derive the following relationship between *VJ* and ε

$$\varepsilon = 6\nu \left(\frac{4\pi DC^* U\,VJ}{Q} \right)^2 \tag{10}$$

For typical values of copepod swimming speed and trail length (Table 1) a value of $VJ = 1$ corresponds to a dissipation rate of $\varepsilon < 5 \times 10^{-9}$ W/kg which is calm water but not unrealistically calm below the pycnocline. A value of $VJ = 100$ corresponds to dissipation rates of $\varepsilon < 5 \times 10^{-5}$ W/kg which is a quite high dissipation rate that is seldom observed in stratified waters below the wave influenced surface layers. Faster swimming copepods have a large advantage in trail detection in turbulent water since $VJ = 1$ corresponds to $\varepsilon = 2 \times 10^{-7}$ W/kg for a fast swimming copepod ($U = 0.01$ m/s) and $\varepsilon = 5 \times 10^{-10}$ W/kg for a slow copepod ($U = 0.0005$ m/s).

The value of $z/z_0 = UT/z_0$ (trail detection distance divided by trail length in the absence of turbulence, T is the detection time scale given in Equation (7)) is shown in Figure 4 for a tidal stratified flow over the Oslo fjord sill using the typical values from Table 1 and dissipation rates estimated from observations with a semi free-falling microstructure shear probe [24]. Seuront and colleagues [7] suggest that trail following is unlikely in the upper ocean. Trails are indeed shortened by >80% in the rather extreme turbulence behind the sill, but there are also calm layers where the trails are only moderately influenced by turbulence.

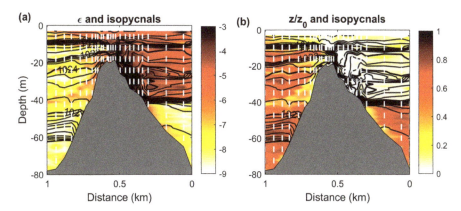

Figure 4. Transects over the Drøbak sill in Oslo fjord of (**a**) dissipation rates of turbulent kinetic energy ($\log_{10}$(W/kg)) [24] and (**b**) trail detection distance relative to that in calm water as functions of distance over the sill and depth for a copepods with swimming speed 0.0015 m/s and detection limit 1 nM. Black contours are lines of constant density (kg m^{-3}), white hatched lines indicate the sampling locations. The transect is performed during spring tide with a strong current from left to right pushing dense water over the sill that cascades down on the lee side and jumps back up again creating strong turbulence.

3.3. Background Concentration

Even for strong turbulence, the constant supply of copepodamides from copepods will cause an increase in background concentration, which when sufficiently large will be balanced by the decay rate of the solute. These background concentrations reach bioactive concentrations in the upper ocean that trigger defensive traits such as harmful algal toxin production [9]. Since zooplankton migrate to deeper waters during daytime to avoid visual predators, the concentration will oscillate. To understand what effective concentration that results up we need to calculate both the time scale, and the equilibrium concentration. The equilibrium between supply and decay can be written as

$$nQ = k\langle C \rangle \qquad (11)$$

where <C> is the average concentration at equilibrium, which can be written as

$$\langle C \rangle = \frac{nQ}{k} \qquad (12)$$

The time scale to reach this equilibrium is k^{-1}. For typical values (Table 1) the equilibrium concentration is ~30 pM which is reached in a time scale of ~5 h. This is an order of magnitude higher than the highest concentrations found in situ [9] which suggest that copepods do not reside sufficiently long in the same water package, or that other loss factors such as vertical mixing and sedimentation removes copepodamides. Alternatively, the production rate of copepodamides reported for *Calanus* sp. [8] may not be representative for other copepods and temperatures.

3.4. Effect of Near Surface Anisotropic Turbulence on Trail Formation

Figure 5 shows snapshots of the stream-wise and vertical velocity. The horizontal velocity is higher than the vertical at this depth and flow situation. The variability is also higher in the horizontal direction, $u_{rms}/w_{rms} \approx 2$. Similar variability ratios are found in channel flow experiments for wind-driven turbulence [14]. This ratio increases with increasing wind stress and vicinity to the surface. This anisotropy distributes signals more horizontally than vertically (Figure 6, Supplementary Video S1). The vertical spread of the plume is, beside turbulence and diffusion, enhanced by the falling sheets of fluid. The mean concentration field is not fully converged due to the limited sampling time which resulted from computational resource limitations and too avoid interference from adjacent sources. However, it is possible to conclude that a vertical search pattern will be more rewarding than a horizontal due to the anisotropy at this depth. The resolution of the model sets a limit on how low signal molecular diffusivities that can be resolved, which gives that the diffusivity used is representative for heat rather than solutes and therefore too high. The signal spread is determined by the diffusion and advection, where a high molecular diffusivity leads to more pronounced diffusion compared to a low diffusivity. Here, advection drives the anisotropic spread whereas the diffusion works towards a more uniform distribution. The anisotropic spread may consequently be slightly more pronounced for copepodamides due to their lower molecular diffusivity. The model can therefore be seen as a conservative approach regarding the anisotropic spread. The copepods are fixed in these simulations, which means that the trail is formed by an intermittent flow, however with a mean flow of 1 mm s^{-1}, past the copepod. In reality the trails will be more continuous due to a more constant velocity of the copepods relative to surrounding water. Supplementary Video S1 shows an animation of the spread around point sources.

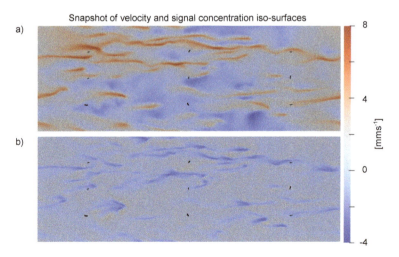

Figure 5. Flow velocity in the plane where the signal is seeded to the domain. Note the same signal scale close to the maximum scale for the stream wise velocity. The nine positions can be seen as the black dots (iso-surfaces of signal concentrations). (**a**) Snapshot ($t = t_0 + 40$ s) of streamwise velocity. (**b**) Snapshot ($t = t_0 + 40$ s) of vertical velocity.

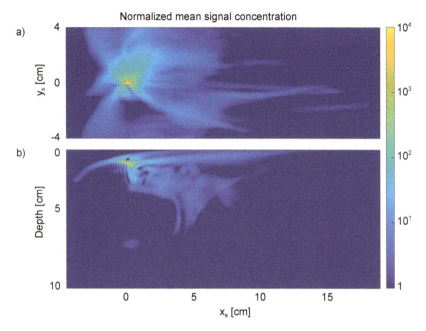

Figure 6. Normalized mean signal concentration. Values below one (dark blue) means that the concentration is less than the total average signal concentration. $x_s = y_s = 0$ and depth = 1 cm define the signal source position. (**a**) Horizontal plane at the seeding depth. (**b**) Vertical plane through the seeding position.

Surfactants, surface-active chemical agents at the ocean-atmosphere of water–air interface, may influence the turbulence in the vicinity of the surface [15,25] and might also attract amphiphilic compounds like the copepodamides and hence serve as a sink for signaling compounds. In addition, the surfactant typically decreases turbulence close to the surface as compared to a clean surface [15].

4. Conclusions

We apply new empirical data on signal substance properties on existing theories on the shortening of trail detection distance by turbulence. This reveals that trails of small copepods are affected by turbulence at quite low dissipation rates of turbulent kinetic energy, whereas the trails of large copepods are unaffected at moderate dissipation rates, which are frequently occurring in calm weather or below the wind mixed layer of stratified natural water. Copepods appear to be aware and maximize encounter rates by forming near surface swarms on quiet days when turbulence levels allow trail following. The anisotropic turbulence in this part of the ocean leads to more horizontal than vertical distribution of cues. Copepods may consequently benefit from adopting more vertical oriented search strategies to encounter chemical trails in this part of the ocean. Moreover, the timescale needed to reach steady state concentrations where exudation is balanced by degradation is in the order of 5 h, which suggests that cue concentration in surface water will oscillate with zooplankton diurnal migrations in a sinusoidal way.

Supplementary Materials: The following are available online at http://www.mdpi.com/2311-5521/5/2/54/s1, Video S1: Solute distribution around point sources in anisotropic turbulence.

Author Contributions: Conceptualization, E.S., S.T.F. and L.A.; methodology, E.S., S.T.F. and L.A.; writing—original draft preparation, E.S., S.T.F. and L.A.; writing—review and editing, E.S., S.T.F. and L.A.; visualization, S.T.F. and L.A.; funding acquisition, E.S., S.T.F. and L.A. All authors have read and agreed to the published version of the manuscript.

Funding: This research was funded by Swedish Research Council VR, grant numbers 2015-05491 and 2019-05238 (to ES). CoCliME which is part of ERA4CS, an ERA-NET initiated by JPI Climate, and funded by EPA (IE), ANR (FR), DLR (DE), UEFISCDI (RO), RCN (NO) and FORMAS (SE), with co-funding by the European Union (Grant 690462).

Conflicts of Interest: The authors declare no conflict of interest. The funders had no role in the design of the study; in the collection, analyses, or interpretation of data; in the writing of the manuscript, or in the decision to publish the results.

Appendix A Calculation of Normalized Mean Signal Concentration

The spatial and temporal averaged scalar concentrations are calculated as follows. The scalar is added to the domain from time t_0 until $t = t_0 + 100$ s. The sampling is done between at $t_1 = t_0 + 40$ s to $t_2 = t_0 + 100$ s with 2.5 s between each realization (snapshot). There are nine positions and for each time realization first the differential concentration is calculated as

$$\Delta S^m_{ijk,t} = S^m_{ijk} - S_t \tag{A1}$$

where m is the position number, ijk are the indices in the x-, j-, and z-direction for that subdomain, t is the time of the realization and S_t is the total amount in the whole domain during that realization. Then the spatial and temporal average for a scalar domain is found by

$$\Delta \bar{\bar{S}}_{ijk} = \frac{\sum_{m=1}^{9} \sum_{t=t_1}^{t_2} \Delta S^m_{ijk,t}}{t \cdot m} \tag{A2}$$

that in turn are normalized using the total amount of scalar at the last realization S_{t_2} as

$$\Delta \widetilde{\bar{\bar{S}}}_{ijk} = \Delta \bar{\bar{S}}_{ijk} / S_{t_2} \tag{A3}$$

References

1. Kiørboe, T. *A Mechanistic Aproach to Plankton Ecology*; Princeton University Press: Princeton, NJ, USA, 2008; p. 228.
2. Berg, H.C. *Random Walks in Biology*; Princeton university press: Princeton, NJ, USA, 1993; p. 152.
3. Bagoien, E.; Kiørboe, T. Blind dating-mate finding in planktonic copepods. I. Tracking the pheromone trail of *Centropages typicus*. *Mar. Ecol. Prog. Ser.* **2005**, *300*, 105–115. [CrossRef]
4. Visser, A.W.; Jackson, G.A. Characteristics of the chemical plume behind a sinking particle in a turbulent water column. *Mar. Ecol. Prog. Ser.* **2004**, *283*, 55–71. [CrossRef]
5. Yen, J.; Lasley, R. *Chemical Communication between Copepods: Finding the Mate in a Fluid Environment*; Springer: New York, NY, USA, 2011; pp. 177–197.
6. Heuschele, J.; Selander, E. The chemical ecology of copepods. *J. Plankton Res.* **2014**, *36*, 895–913. [CrossRef]
7. Seuront, L.; Stanley, H.E. Anomalous diffusion and multifractality enhance mating encounters in the ocean. *Proc. Natl. Acad. Sci. USA* **2014**, *111*, 2206–2211. [CrossRef] [PubMed]
8. Selander, E.; Kubanek, J.; Hamberg, M.; Andersson, M.X.; Cervin, G.; Pavia, H. Predator lipids induce paralytic shellfish toxins in bloom-forming algae. *Proc. Natl. Acad. Sci. USA* **2015**, *112*, 6395–6400. [CrossRef] [PubMed]
9. Selander, E.; Berglund, E.; Engström, P.; Berggren, F.; Eklund, J.; Harðardóttir, S.; Lundholm, N.; Grebner, W.; Andersson, M. Copepods drive large-scale trait-mediated effects in marine plankton. *Sci. Adv.* **2019**, *5*, eaat5096. [CrossRef] [PubMed]
10. Selander, E.; Jakobsen, H.H.; Lombard, F.; Kiørboe, T. Grazer cues induce stealth behavior in marine dinoflagellates. *Proc. Natl. Acad. Sci. USA* **2011**, *108*, 4030–4034. [CrossRef] [PubMed]
11. Byron, E.; Whitman, P.; Goldman, C. Observations of copepod swarms in Lake Tahoe 1. *Limnol. Oceanogr.* **1983**, *28*, 378–382. [CrossRef]
12. Fredriksson, S.; Arneborg, L.; Nilsson, H.; Handler, R. Surface shear stress dependence of gas transfer velocity parameterizations using DNS. *J. Geophys. Res. Ocean.* **2016**, *121*, 7369–7389. [CrossRef]

13. Fredriksson, S.T.; Arneborg, L.; Nilsson, H.; Zhang, Q.; Handler, R.A. An evaluation of gas transfer velocity parameterizations during natural convection using DNS. *J. Geophys. Res. Ocean.* **2016**, *121*, 1400–1423. [CrossRef]
14. Kim, J.; Moin, P.; Moser, R. Turbulence statistics in fully developed channel flow at low Reynolds number. *J. Fluid Mech.* **1987**, *177*, 133–166. [CrossRef]
15. Zhang, Q.; Handler, R.A.; Fredriksson, S.T. Direct numerical simulation of turbulent free convection in the presence of a surfactant. *Int. J. Heat Mass Transf.* **2013**, *61*, 82–93. [CrossRef]
16. Csanady, G.T. *Turbulent Diffusion in the Environment*; Springer Science & Business Media: Berlin/Heidelberg, Germany, 2012; Volume 3.
17. Jackson, G.A.; Kiørboe, T. Zooplankton use of chemodetection to find and eat particles. *Mar. Ecol. Prog. Ser.* **2004**, *269*, 153–162. [CrossRef]
18. Batchelor, G.K. Small-scale variation of convected quantities like temperature in turbulent fluid Part 1. General discussion and the case of small conductivity. *J. Fluid Mech.* **1959**, *5*, 113–133. [CrossRef]
19. Tiselius, P.; Jonsson, P.R. Foraging Behavior of 6 Calanoid Copepods-Observations and Hydrodynamic Analysis. *Mar. Ecol. Prog. Ser.* **1990**, *66*, 23–33. [CrossRef]
20. Arias, A.; Selander, E.; Saiz, E.; Calbet, A. Predator chemical cues effects on the diel feeding behaviour of marine protists. *Proc. R. Acad. Sci. Proc. B Biol. Sci.*. (in press).
21. Evans, R.; Dal Poggetto, G.; Nilsson, M.; Morris, G.A. Improving the interpretation of small molecule diffusion coefficients. *Anal. Chem.* **2018**, *90*, 3987–3994. [CrossRef] [PubMed]
22. Thorpe, S.A. *An Introduction to Ocean Turbulence*; Cambridge University Press: Cambridge, UK, 2007.
23. Csanady, G.T. *Air-Sea Interaction: Laws and Mechanisms*; Cambridge University Press: Cambridge, UK, 2001.
24. Staalstrøm, A.; Arneborg, L.; Liljebladh, B.; Broström, G. Observations of turbulence caused by a combination of tides and mean baroclinic flow over a fjord sill. *J. Phys. Oceanogr.* **2015**, *45*, 355–368. [CrossRef]
25. McKenna, S.; McGillis, W. The role of free-surface turbulence and surfactants in air–water gas transfer. *Int. J. Heat Mass Transf.* **2004**, *47*, 539–553. [CrossRef]

© 2020 by the authors. Licensee MDPI, Basel, Switzerland. This article is an open access article distributed under the terms and conditions of the Creative Commons Attribution (CC BY) license (http://creativecommons.org/licenses/by/4.0/).

Review

Numerical Simulations of Flow around Copepods: Challenges and Future Directions

Iman Borazjani

J. Mike Walker '66 Department of Mechanical Engineering, Texas A&M University, College Station, TX 77843, USA; iman@tamu.edu; Tel.: +1-979-458-578

Received: 26 February 2020; Accepted: 14 April 2020; Published: 17 April 2020

Abstract: Copepods are small aquatic creatures which are abundant in oceans as a major food source for fish, thereby playing a vital role in marine ecology. Because of their role in the food chain, copepods have been subject to intense research through different perspectives from anatomy, form-function biology, to ecology. Numerical simulations can uniquely support such investigations by quantifying: (i) the force and flow generated by different parts of the body, thereby clarify the form-function relation of each part; (ii) the relation between the small-scale flow around animal and the large-scale (e.g., oceanic) flow of its surroundings; and (iii) the flow and its energetics, thereby answering ecological questions, particularly, the three major survival tasks, i.e., feeding, predator avoidance, and mate-finding. Nevertheless, such numerical simulations need to overcome challenges involving complex anatomic shape of copepods, multiple moving appendages, resolving different scales (appendage-, animal- to large-scale). The numerical methods capable of handling such problems and some recent simulations are reviewed. At the end, future developments necessary to simulate copepods from animal- to surrounding-scale are discussed.

Keywords: copepod; plankton; numerical simulation; immersed boundary method; multi-scale simulations; form-function relation

---

1. Introduction

Copepods (from the Greek word for *oar feet*) are among the most diverse animals in the world with more than 14,000 described species. They can be found in almost any kind of aquatic environment, from subzero waters of Arctic to hot springs, from the top of Himalayas to 10,000 meters down the deep sea, in mud, subterranean groundwater, lakes, seas and oceans [1]. Some copepods are parasitic and live off other animals such as fish—e.g., salmon louse *Lepeophtheirus salmonis* Kroyer is a devastating pest for salmon farms [1]. They are among the most abundant animals on the planet and a major food source for fish. They play a vital role in the food chain and marine ecology which have made them the subject of intense research through different perspectives from anatomy, form-function biology, to ecology. In this manuscript, the numerical methods capable of supporting such research are reviewed and their limitations and strengths are discussed.

The anatomy of copepods is well-studied and a general body plan is consistent across myriad orders of these small crustaceans. Their size typically ranges between 0.5 to 5 mm. They have a segmented body with several appendages attached to it (see Figure 1). A copepod has a pair of first and a pair of second antennae, the first pair is long and the other is short, mandibular palps, two pairs of maxillas, and a pair of maxilliped. There are four or five pairs of swimming legs attached to the abdomen and the urosome with setae at the caudal area. They have one simple eye in the middle of the head (at least in the larval stage), which can only tell the difference between light and dark. Most of the sensory ability of copepods comes from the chemoreceptors on the mouthparts [2] and mechanoreceptors over appendages especially the first antennae [3]. Overall, simulating a copepod with an anatomically realistic shape and moving appendages poses a great challenge even to the most advanced numerical tools—see Section 2.

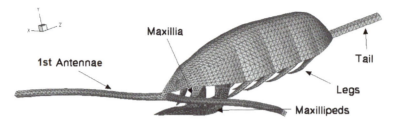

Figure 1. A typical calanoid copepod with all the appendages modeled in computer aided design (CAD) and meshed with triangular elements for numerical simulations. Reproduced with permission from [4].

The behavior of copepods and the function of their appendages has been typically investigated through observations using high-speed cameras. Copepods feed on a wide variety of prey, ranging from algae of a few micrometers to metazooplankton and fish larvae [5]. They use their mouth appendages i.e., the antennae, maxillia and maxillipeds to create a feeding current toward the mandibular palps (jaws) [6–8]. The flow field created by the feeding copepod is laminar with a low Reynolds number based on length and swimming speed of order 1 to 10, where the viscous forces are dominant. Copepods spend most of their time swimming and feeding at this low Reynolds number environment. However, if a threat is detected they respond with rapid jumps to escape from harm's way. Such escape maneuvers give copepods a much greater chance of survival than the zooplanktons which do not exhibit such predator-avoidance mechanism [9]. The Reynolds number during jumps may reach as high as several thousands, which may even transition to turbulence.

The appendages move differently during an escape maneuver from the normal cruising mode. When cruising, only the mouth appendages move to create a laminar feeding current while in the escape the power strokes of antennae and the legs are mainly used to create large acceleration. The power-stroke starts with the beating of the first antennae followed by multiple metachronal beating of the legs while other mouth appendages stay in the retracted position [10]. During the return stroke swimming legs minimize their surface area and move synchronously to the original position [10–12]. Such power-strokes can give copepods the incredible speed of 50 to 500 body-length per second during escape [12–14]. To understand how copepods achieve such high velocities, the hydrodynamic forces produced by the appendages movements should be determined.

A few experimental studies have been conducted to measure the hydrodynamic forces during the power and return strokes. Alcaraz and Strickler [11] examined the relation of forces and appendages movement for escape mode by measuring the spring force attached to the tethered copepod while filming the appendages movement from the side. They found the force to be in the thrust (forward) direction during the power stroke and in the drag (backward) direction during the return stroke. More recently, Lenz et al. [12] performed a similar study. They reported that the peaks on the force record corresponded to the power stroke of each leg and the stroke of forth and third pairs of legs produced the largest peak. The estimated power per muscle mass by Lenz et al. [12] was found to be particularly high relative to other organisms. In the above studies, the force record was the total force produced by all the appendages and it was not completely clear how much each appendage contributed to the total force. A major contribution of numerical simulations is that they specifically provide how much each appendage contributes to the total force and how the appendages movements affect the forces produced, thereby clarify the form-function relation of anatomical structures or appendages.

Another point of interest in copepod research is the flow field that is generated by the copepods and how copepods respond to the flow in their environment [15,16]. In fact, copepods respond to hydrodynamic signals within a few milliseconds [17]. They use their mechanoreceptors to feel the hydrodynamic disturbances in the flow [18]. It is believed that they can distinguish between prey, predator, or mate from the hydrodynamic signatures of the flow [19]. In addition, the hydroyanmic disturbance that they create is used by fish to capture them. The flow field and the coherent vortical structures has been visualized in experiments with live copepods [14,19–22]. Such flow visualizations typically capture the flow over a 2D plane. 3D flow measurements are possible, e.g., using tomographic particle image velocimetry (tomo-PIV) [23], but they are more challenging and normally have a lower resolution than 2D PIV. Numerical simulations can complement such experiments by providing the detailed 3D flow around the copepods. Nevertheless, the numerical simulation of a copepod is a challenging undertaking due to its complex shape and the thin appendages and their rapid movements. In the next section, available methods for simulating flow around copepods with different levels of complexity are reviewed and their strengths and limitations are discussed.

2. Overview of Numerical Methods for Simulations of Copepods

The methods for simulating copepods can be classified into two main categories: Simulations that resolve the shape and motion of copepod (appendage-scale methods), and the simulations that ignore the exact shape of a copepod or the motion of its appendages and model their effect as a force field on the flow (force-field method). The former is better suited for investigating form-function relation of different anatomical features of a copepod, whereas the latter is useful for investigating the general flow features generated by copepod (copepod-scale), the interaction of copepod-scale flow and the large-scale (e.g., oceanic) flow, or their collective behavior. In what follows, these two categories are further discussed.

2.1. Force-Field Simulations

When the large- or copepod-scale flow is of interest, i.e., the length scales of surrounding flow are much larger than the appendages size, the copepod shape or its appendages are ignored but its effect on the flow is modeled as a force field—see the reviews [24,25] for more details and applications of this method. In fact, the background flow is governed by the incompressible Navier-Stokes (or Stokes depending on the Reynolds number of the flow) equations and a force is added at the locations where the copepods are present [26]. This method is similar to what has been used for simulation of small particles [27] or bacteria and other microorganisms (active fluids) in the flow [28].

The 3D incompressible, non-dimensional Navier-Stokes equations are as follows:

$$\nabla \cdot \mathbf{u} = 0$$
$$\frac{D\mathbf{u}}{Dt} = -\nabla p + \frac{1}{Re}\nabla^2 \mathbf{u} + \mathbf{f} \quad (1)$$

where $D/Dt = \partial/\partial t + \mathbf{u}.\nabla$, $\mathbf{u}$ is the velocity vector, p is pressure, Re is the Reynolds number, and $\mathbf{f}$ is the force field (force per unit volume) exerted by the copepods onto the flow. The above equations require appropriate boundary conditions to form a well-posed problem. The boundary conditions are the no-slip condition at copepod locations in flow

$$\mathbf{u} = \mathbf{V}_{swimming} \quad (2)$$

where $\mathbf{V}_{swimming}$ is the swimming velocity of the copepod.

This method has been used to simulate the flow at the copepod-scale. Jiang et al. [26] developed five models for the force $\mathbf{f}$ in Stokes flow for different types of steady swimming such as hovering, freely sinking, and upward/backward/forward swimming. The force field $\mathbf{f}$ was obtained through force balance by assuming the shape of the copepod body as a sphere and using Stokes drag formulas. More recently, Jiang and Kiørboe [29] used an impulsive stresslet to model the force field generated by a jumping copepod. In addition, Jiang and Kiørboe [30] simulated the copepod jumps by assuming an spheroidal shape for the body and applying a force field to account for the beating legs. The flow field created by self-propelled jumping copepod compared well with PIV measured flow fields and spatial decay rate [30].

The force field method can also be used to simulate the interaction of copepod with the large-scale flow, e.g., turbulence. In the pioneering work of Yamazaki et al. [31], direct numerical simulations of an isotropic turbulence were performed and copepods were simulated as passively drifting (inertialess) particles by the turbulent flow plus a random walk. Later, Squires and Yamazaki [32] simulated marine particles with inertia in a similar setup and showed the preferential concentration (rather than a homogeneous distribution) generated by such particles. Lewis and Pedley [33] used a probablity density function of predator and prey velocities, rather than direct numerical simulations, to model the contact of microorganisms in turbulent flow. Recently, Ardeshiri et al. [34] carried out direct numerical simulations of isotropic turbulence and simulated the motion of copepods as particle similar to Yamazaki et al. [31] but instead of the random walk they developed and used a Lagrangian model based on the recorded velocities of copepods in an experimental setting. In all of the above studies [31,32,34], a non-uniform body force was applied to the largest scales of motion in order to maintain a statistically stationary flow, but the effect of copepods on the turbulent flow was ignored and no force from the copepods were added to the right hand side of Equation (1). In fact, the copepods' motion were determined by the flow but the existence of the copepod did not affect the flow, i.e., the copepods' motion was coupled one-way to the flow. Nevertheless, it is known that the existence of particles in the turbulent flow affects the turbulent characteristics as well as particle dispersion, which requires two-way coupling [35,36]. In two-way coupling, the force exerted by the copepods on the flow needs to be added to the right hand side of Equation (1).

2.2. Appendage-Scale Simulations

For simulations of copepods with realistic body and appendages shapes, the Navier-Stokes Equation (1) are solved over a grid with appropriate no-slip conditions on the body (Equation (2)). Nevertheless, handling the motion of copepod appendages require special techniques because the boundary at which the no-slip conditions should be applied is moving. The numerical methods for moving boundary problems are classified into two main categories [37]: (a) boundary-conforming methods in which the grid moves the moving boundary and (b) non boundary-conforming methods in which the grid is fixed and the moving boundary moves over a fixed background mesh.

In boundary-conforming methods, also known as the arbitray Lagrangian-Eulerian (ALE) methods [38], the Navier-Stokes Equation (1) are modified to account for the grid motion. The ALE methods can efficiently keep a high-resolution near the moving boundaries. For large deformations, however, the grid can become highly skewed or stretched, i.e., the grid quality degrades, which may cause convergence issues for the numerical method. Remeshing may solve the grid quality, but it is computationally expensive and difficult.

Non boundary-conforming, i.e., fixed-grid, methods do not need any modification to the Navier-Stokes equations as the grid is fixed and does not move with the boundary. These methods are more suitable for large deformations because they do not create highly skewed grids [39]. However, additional complexities arise in terms of identifying the grid nodes adjacent to the moving boundaries and transferring the effects onto those nodes. There are different fixed-grid methods, e.g., the immersed boundary method [39], cut-cell methods [40], and fictitious domain method [41], among others. Out of these methods, only the immersed boundary has been used for simulations of copepods [4]. Consequently, only the immersed boundary is briefly described below.

The original immersed boundary, pioneered by Peskin [42–44], smeared the effect of boundaries over several grid nodes, which required additional resolution near the boundary. Since then, a number of sharp-interface methods have been developed [39]. In the method used for copepods [4], the sharp-interface is maintained by reconstructing the boundary conditions at the nodes that are exterior to, but adjacent to the immersed-boundary surface using a quadratic interpolation along the local normal to the boundary [45]. The background nodes at each time step are classified into fluid, wall, and immersed boundary nodes using an efficient ray-tracing algorithm [46]. The method uses curvilinear grids as the background grid to efficiently stretch the grid in regions of interest, i.e, the curvilinear-immersed boundary (CURVIB) method [47]. The method is fully parallelized using Message Passing Interface (MPI) and Portable, Extensible Toolkit for Scientific Computation (PETSc) [48]. It has been fully validated [46,49–51] and applied to a wide range of applications: vortex induced vibrations [52,53], aquatic locomotion [54–63], cardiovascular flows [64–70], sediment transport [71], large-eddy simulations and flow control [72–75], rheology [76,77], and copepods [4]. In the next section, some of the copepod simulations using this method are presented and discussed.

3. Results from Previous Simulations of Copepods

The earliest numerical simulations of copepods considered copepods as particles drifting by a turbulent flow [31,32]. In their pioneering work Yamazaki et al. [31] and Squire and Yamazaki [32] tracked the trajectories of 512 and 165,888 copepods (particles), respectively, drifting in an isotropic turbulent flow. The results showed that the contact was increased due to turbulence [31] and clustering can occur with densities of 10 to 60 times of the mean average population density [32]. Recently, Ardeshiri et al.[34] simulated 256,000 copepods with a Lagrangian model of jumping behavior in an isotropic turbulent flow and found clustering of jumping copepods. Furthermore, they used their simulations to estimate the contact rate of jumping copepods and found that it can be increased by a factor up to 10^2 compared to that experienced by passively transported fluid tracers of the same size [78]. These simulations investigated the role of large-scale flow on the copepod motion, but ignored the copepod-scale flow generated by the copepods.

The first attempt to simulate copepod swimming numerically was reported in a series of papers by Jiang et al [26,79–81]. In their theoretical work, the force field approach (Section 2.1) was used by approximating the shape of the copepod as a sphere and modeling the force for different swimming conditions [26,29]. In their numerical simulations [79,81], a realistic body shape was used but the multiple moving appendages were neglected and their effect was accounted for via a distribution of body forces acting on the water. The first simulation of an anatomically realistic copepod with all of its swimming appendages was reported by Gilmanov and Sotiropoulos [45] who employed a sharp-interface Cartesian method (Section 2.2) to a model the swimming of a tethered copepod. Later, Borazjani et al. [4] built on that work by carrying out high-resolution simulations over a range of governing parameters with more biologically accurate kinematics prescribed from high-resolution experiments.

Borazjani et al. [4] used 3D numerical simulations of a tethered copepod to investigate the role of the first antennae in production of hydrodynamic force during hopping (Figure 1). The details of body shape and the kinematics of the appendages were prescribed from experimental observations using a dual digital holography setup with 200 frame per second high-speed camera to record the motion of hopping copepods [21]. By analyzing the high-resolution experimental holographic movies frame-by-frame, it was observed that the copepod antennae deform distinctively during the return stroke to the fully open position [4]. Consequently, two sets of simulations were conducted to examine the importance of the realistic motion of the antennae on the hydrodynamics of during hopping as observed from the top view in Figure 2 [4]: one treating each antenna as a rigid, oar-like structure moving symmetrically during power and return stroke; and the other using an asymmetrical motion with a deformed shape during the return stroke which was prescribed from experiments.

The force produced by each individual appendage was computed directly from the numerical simulations and its relative significance on the total hydrodynamic force was quantified [4]. The computed results: (1) show that for both cases the antennae are major contributors to the net thrust during hopping; and (2) clearly demonstrate the significant hydrodynamic benefit in terms of thrust enhancement and drag reduction due to biologically realistic, asymmetric antenna motion. In addition, the antennae were found to produce the largest drag- and thrust-type forces among all appendages regardless of whether the antennae were treated as rigid or deformable. This conclusion, however, could be partly affected by the fact that the leg kinematics in the present model does not exhibit the required degree of biological realism. Nevertheless, the large drag of the antennae provides a plausible explanation why copepods have been observed to use multiple leg strokes before re-deploying their antennae. Finally, the calculated net force time history over a swimming cycle explains why copepods swimming in the ocean move in bursts (jumps) during escape. The high thrust (produced by all appendages) during the power stroke, which rapidly accelerates the copepod, is followed by a large drag force (produced by the body and the appendages) during the return stroke, which decelerates the copepod, yielding an intermittent, burst-like behavior during the swimming cycle.

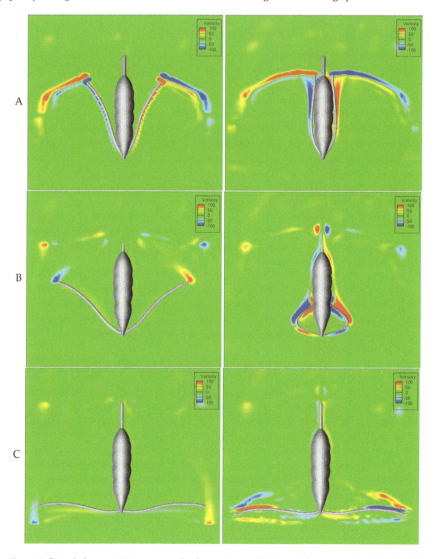

Figure 2. Out-of-plane vorticity contours for the top view mid plane of the copepod at (**A**) t/T = 0.33 (**B**) t/T = 0.66 (**C**) t/T = 1 for the rigid antennae (**left**) and deformable antennae (**right**) (Re = 300, T = 0.3). Reproduced with permission from [4]. See Video S1 in supplementary materials for a movie of the deformable antennae.

The numerical simulations also provide the 3D flow around the copepod. The antennae create complex vortical structures that can be viewed from the top (Figure 2). Figure 2A shows the antennae at the end of the its power stroke, which creates a strong shear layer at the tips of each antenna. Figure 2B,C shows the wake at the middle and end of the return stroke, respectively. It can be observed that the downstream wake structures have been destroyed by the rigid antennae while still moving downstream in the deformable case i.e., more thrust-type wake for the deformable antennae. The numerical simulations also elucidate for the first time the origin of the distinct toroidal vortices observed in flow visualization experiments in the wake of escaping copepods [19]. Visualizing the coherent structures by the iso-surfaces of q-criterion (defined as the difference between the norm of vorticity and strain-rate [82]; consequently, q> 0 are regions where the rotation rate dominates the strain rate, i.e., a vortex) in the wake of the copepod showed the three-dimensional structure of the toroidal vortices (Figure 3). It was found that the toroidal vortices are formed mainly from the leg strokes by forming two columnar vortices, which are attached to each other at the end forming an arch-like vortex loop [4].

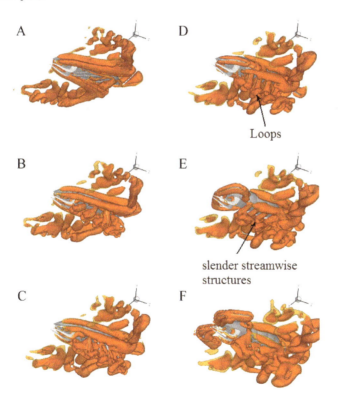

Figure 3. Vortical structures visualized by the iso-surfaces of q-criterion around the tethered copepod with deformable antennae ($Re = 300$, $T = 0.3$) at different time instance (**A**) $t/T = 0.28$ the first stoke of legs begins by the last pair (**B**) $t/T = 0.36$ the middle of legs power-strokes (**C**) $t/T = 0.48$ almost the end of the legs power-stroke (**D**) $t/T = 0.56$ the legs return-stroke has started (**E**) $t/T = 0.66$ towards the end of the legs return-stroke (**F**) $t/T = 0.86$ the legs return stroke has been ended for some time. Reproduced with permission from [4]. See Video S2 in supplementary materials for a movie of the wake structure.

4. Discussions and Future Directions

Numerical simulations have come a long way from simplified geometries to realistic anatomic ones. Numerical simulations can provide the 3D flow around copepods with realistic anatomies and appendages motion, which is challenging to measure experimentally, and provide hydrodynamic forces generated by individual appendages or body. In addition, one of the main benefits of numerical simulations is to test hypothetical cases which might not be possible in real world. Such numerical simulations can play an important role in investigating the form-function relations of biological features. In fact, Borazjani et al. [4] used numerical simulations to investigate the form and function of copepod antennae during hopping. This was possible because the force generated by each appendage can be computed separately from others in numerical simulations which is quite challenging, if not impossible, to do in experiments. In addition, this was also possible through a comparative analysis between a realistic antenna motion and a hypothetical symmetric motion (not observed in nature).

Appendage- and copepod-scale flow simulations need experiments to provide the accurate anatomical geometries and motions for their results to be biologically relevant. The results of such simulations should be validated against available measurements before applying them to hypothetical cases for comparative analysis. Such complementary simulations and experiments are powerful tools to advance knowledge and test hypotheses on form and function of different biological features or behaviors. There are many possibilities in this line of research as many features of copepods has yet to be simulated. The current simulations were mostly tethered [4] and self-propelled simulations, which have been performed for other aquatic swimmers [55,59,61,62,83], has yet to be carried out for appendage-scale flow of copepods. Such simulations can investigate the role of multiple metachronal beat of legs during multiple hops for feeding or jumps to avoid predators. In addition, the role of hairy appendages versus non-porous ones is an interesting feature to investigate. Nevertheless, simulations with hairy or flexible appendages need: (a) realistic material properties measured experimentally for flexible appendages; and (b) a finite element solver coupled with a fluid solver [49] to solve for the deformations due to hydrodynamic forces.

Apart from investigating the form and function of different biological features or appendages, investigating how the copepods interact with the large-scale surrounding flow is an exciting new avenue for research. It is interesting to see how different flow regimes, e.g., turbulence [15] or vortices [16,84], affect the motion (either cruising or hopping) of copepods. In addition, the effect of vortices or flow gradients on the deformation of hairy and deformable appendages may shed light on the mechanosensory mechaisms of copepods. Appendage- and copepod-scale simulations can help in such studies.

Appendage- and copepod-scale simulations, however, cannot investigate the effect of presence of many copepods on the large-scale surrounding flow. It is well-known that the presence of small microorganisms [28] and even passive particles [77] can affect the large-scale flow and its rheology. Previous simulations of copepods motion in large-scale flow [31,32,34] did not consider the effect of copepods on the large-scale flow, i.e., one way coupled (Section 2.1). To investigate the effect of copepods on the large-scale surrounding flow a multi-scale, two-way coupled simulation is required. One possible way for such multi-scale simulations is to couple appendage-scale (Section 2.2) to large-scale (Section 2.1) simulations. Each appendage-scale simulation will provide the force at the copepod location (**f** in the right-hand-side of Equation (1)) for the large-scale simulation while the large-scale simulation will provide the flow conditions (incoming velocity magnitude and direction) at copepod locations for appendage-scale simulations, i.e., two-way coupled. Note such simulations will be computationally expensive and using parallel computing is a must. Finally, such multi-scale simulations do not need modeling (random walks [31] or Lagrangian modeling of jumps [34]) to calculate the trajectories of copepods' motion in a turbulent flow, thereby providing a more accurate contact rate and clustering of copepods. Contact rates and clustering of copepods are directly related to the probability of encountering food, mates, or predators. Better approximations of contact rate through two-way, multi-scale simulations enable answering important ecological question involving the three major survival tasks, i.e., feeding, predator avoidance, and mate-finding.

Supplementary Materials: The following are available online at http://www.mdpi.com/2311-5521/5/2/52/s1, Video S1: The top view of out-of-plane vorticity contours on the midplane of the copepod with deformable antenna. Video S2: 3D vortical structures visualized by the iso-surfaces of q-criterion around the copepod with deformable antenna.

Funding: This work was partially supported by National Science Foundation career grant CBET 1453982.

Conflicts of Interest: The authors declare no conflict of interest.

References

1. Boxshall, G.A.; Halsey, S.H. *An Introduction to Copepod Diversity*; Ray Society: Andover, UK, 2004.
2. Friedman, M.M.; Strickler, J.R. Chemoreceptors and Feeding in Calanoid Copepods (Arthropoda: Crustacea). *Proc. Natl. Acad. Sci. USA* **1975**, *72*, 4185. [CrossRef] [PubMed]
3. Strickler, J.R.; Bal, A.K. Setae of the First Antennae of the Copepod Cyclops scutifer (Sars): Their Structure and Importance. *Proc. Natl. Acad. Sci. USA* **1973**, *70*, 2656. [CrossRef] [PubMed]
4. Borazjani, I.; Sotiropoulos, F.; Malkiel, E.; Katz, J. On the role of copepod antenna in the production of hydrodynamic force during hopping. *J. Exp. Biol.* **2010**, *213*, 3019–3035. [CrossRef] [PubMed]
5. Nejstgaard, J.C.; Frischer, M.E.; Simonelli, P.; Troedsson, C.; Brakel, M.; Adiyaman, F.; Sazhin, A.F.; Artigas, L.F. Quantitative PCR to estimate copepod feeding. *Mar. Biol.* **2008**, *153*, 565–577. [CrossRef]
6. Koehl, M.A.R.; Strickler, J.R. Copepod Feeding Currents: Food Capture at Low Reynolds Number. *Limnol. Oceanogr.* **1981**, *26*, 1062–1073. [CrossRef]
7. Svetlichny, L.S.; Hubareva, E.S. The energetics of Calanus euxinus: Locomotion, filtration of food and specific dynamic action. *J. Plankton Res.* **2005**, *27*, 671. [CrossRef]
8. van Duren, L.A.; Stamhuis, E.J.; Videler, J.J. Copepod feeding currents: Flow patterns, filtration rates and energetics. *J. Exp. Biol.* **2003**, *206*, 255. [CrossRef]
9. Trager, G.; Achituv, Y.; Genin, A. Effects of prey escape ability, flow speed, and predator feeding mode on zooplankton capture by barnacles. *Mar. Biol.* **1994**, *120*, 251. [CrossRef]
10. Strickler, J.R. Swimming of planktonic Cyclops species (Copepoda, Crustacea): Pattern, movements and their control. In *Swimming and Flying in Nature*; Springer: Boston, MA, USA, 1975; Volume 2.
11. Alcaraz, M.; Strickler, J.R. Locomotion in copepods: Pattern of movements and energetics of Cyclops. *Hydrobiologia* **1988**, *167–168*, 409. [CrossRef]
12. Lenz, P.H.; Hower, A.E.; Hartline, D.K. Force production during pereiopod power strokes in Calanus finmarchicus. *J. Mar. Syst.* **2004**, *49*, 133–144. [CrossRef]
13. Buskey, E.J.; Lenz, P.H.; Hartline, D.K. Escape behavior of planktonic copepods in response to hydrodynamic disturbances: high speed video analysis. *Mar. Ecol. Prog. Ser.* **2002**, *235*, 135–146. [CrossRef]
14. van Duren, L.A.; Videler, J.J. Escape from viscosity: The kinematics and hydrodynamics of copepod foraging and escape swimming. *J. Exp. Biol.* **2003**, *206*, 269. [CrossRef] [PubMed]
15. Yamazaki, H.; Squires, K.D. Comparison of oceanic turbulence and copepod swimming. *Mar. Ecol. Prog. Ser.* **1996**, *144*, 299–301. [CrossRef]
16. Webster, D.; Young, D.; Yen, J. Copepods' response to Burgers' vortex: Deconstructing interactions of copepods with turbulence. *Integr. Compar. Biol.* **2015**, *55*, 706–718. [CrossRef]
17. Lenz, P.H.; Hartline, D.K. Reaction times and force production during escape behavior of a calanoid copepod, Undinula vulgaris. *Mar. Biol.* **1999**, *133*, 249. [CrossRef]
18. Yen, J.; Lenz, P.H.; Gassie, D.V.; Hartline, D.K. Mechanoreception in marine copepods: Electrophysiological studies on the first antennae. *J. Plankton Res.* **1992**, *14*, 495. [CrossRef]
19. Yen, J.; Strickler, J.R. Advertisement and Concealment in the Plankton: What Makes a Copepod Hydrodynamically Conspicuous? *Invertebr. Biol.* **1996**, *115*, 191–205. [CrossRef]
20. Catton, K.B.; Yen, J.; Webster, D.R.; Brown, J. Quantitative analysis of tethered and free-swimming copepodid flow fields. *J. Exp. Biol.* **2007**, *210*, 299–310. [CrossRef] [PubMed]
21. Malkiel, E.; Sheng, J.; Katz, J.; Strickler, J.R. The three-dimensional flow field generated by a feeding calanoid copepod measured using digital holography. *J. Exp. Biol.* **2003**, *206*, 3657–3666. [CrossRef] [PubMed]

22. Stamhuis, E.; Videler, J. Quantitative flow analysis around aquatic animals using laser sheet particle image velocimetry. *J. Exp. Biol.* **1995**, *198*, 283–294. [PubMed]
23. Murphy, D.; Webster, D.; Yen, J. A high-speed tomographic PIV system for measuring zooplanktonic flow. *Limnol. Oceanogr. Methods* **2012**, *10*, 1096–1112. [CrossRef]
24. Jiang, H.; Osborn, T.R. Hydrodynamics of Copepods: A Review. *Surv. Geophys.* **2004**, *25*, 339. [CrossRef]
25. Jiang, H. Numerical simulation of the flow field at the scale size of an individual copepod. In *Handbook of Scaling Methods in Aquatic Ecology: Measurement, Analysis, Simulation*; CRC Press: Boca Raton, FL, USA, 2004; pp. 333–359.
26. Jiang, H.; Osborn, T.R.; Meneveau, C. The flow field around a freely swimming copepod in steady motion. Part I: Theoretical analysis. *J. Plankton Res.* **2002**, *24*, 167. [CrossRef]
27. Zhong, W.; Yu, A.; Liu, X.; Tong, Z.; Zhang, H. DEM/CFD-DEM Modelling of Non-spherical Particulate Systems: Theoretical Developments and Applications. *Powder Technol.* **2016**, *302*, 108–152. [CrossRef]
28. Koch, D.L.; Subramanian, G. Collective hydrodynamics of swimming microorganisms: Living fluids. *Annu. Rev. Fluid Mech.* **2011**, *43*, 637–659. [CrossRef]
29. Jiang, H.; Kiørboe, T. The fluid dynamics of swimming by jumping in copepods. *J. R. Soc. Interface* **2011**, *8*, 1090–1103. [CrossRef]
30. Jiang, H.; Kiørboe, T. Propulsion efficiency and imposed flow fields of a copepod jump. *J. Exp. Biol.* **2011**, *214*, 476–486. [CrossRef]
31. Yamazaki, H.; Osborn, T.R.; Squires, K.D. Direct numerical simulation of planktonic contact in turbulent flow. *J. Plankton Res.* **1991**, *13*, 629–643. [CrossRef]
32. Squires, K.D.; Yamazaki, H. Preferential concentration of marine particles in isotropic turbulence. *Deep Sea Res. Part I Oceanogr. Res. Pap.* **1995**, *42*, 1989–2004. [CrossRef]
33. Lewis, D.; Pedley, T. Planktonic contact rates in homogeneous isotropic turbulence: Theoretical predictions and kinematic simulations. *J. Theor. Biol.* **2000**, *205*, 377–408. [CrossRef]
34. Ardeshiri, H.; Benkeddad, I.; Schmitt, F.G.; Souissi, S.; Toschi, F.; Calzavarini, E. Lagrangian model of copepod dynamics: Clustering by escape jumps in turbulence. *Phys. Rev. E* **2016**, *93*, 043117. [CrossRef] [PubMed]
35. Elghobashi, S.; Truesdell, G. On the two-way interaction between homogeneous turbulence and dispersed solid particles. I: Turbulence modification. *Phys. Fluids A Fluid Dyn.* **1993**, *5*, 1790–1801. [CrossRef]
36. Truesdell, G.; Elghobashi, S. On the two-way interaction between homogeneous turbulence and dispersed solid particles. II. Particle dispersion. *Phys. Fluids* **1994**, *6*, 1405–1407. [CrossRef]
37. Borazjani, I.; Akbarzadeh, A. Large Eddy Simulations of Flows with Moving Boundaries. In *Modeling and Simulation of Turbulent Mixing and Reaction*; Springer: Singapore, 2020; pp. 201–225.
38. Donea, J.; Huerta, A.; Ponthot, J.P.; Rodríguez-Ferran, A., Arbitrary Lagrangian–Eulerian Methods. In *Encyclopedia of Computational Mechanics*; American Cancer Society: Atlanta, GA, USA, 2004; Chapter 14. [CrossRef]
39. Mittal, R.; Iaccarino, G. Immersed boundary methods. *Annu. Rev. Fluid Mech.* **2005**, *37*, 239–261. [CrossRef]
40. Tucker, P.; Pan, Z. A Cartesian cut cell method for incompressible viscous flow. *Appl. Math. Model.* **2000**, *24*, 591–606. [CrossRef]
41. Glowinski, R.; Pan, T.W.; Hesla, T.I.; Joseph, D.D. A distributed Lagrange multiplier/fictitious domain method for particulate flows. *Int. J. Multiphase Flow* **1999**, *25*, 755–794. [CrossRef]
42. Peskin, C. Numerical Analysis of Blood Flow in the Heart. *J. Comput. Phys.* **1977**, *25*, 220. [CrossRef]
43. Peskin, C.; McQueen, D. A three-dimensional computational method for blood flow in the heart. 1. Immersed elastic fibers in a viscous incompressible fluid. *J. Comput. Phys.* **1989**, *81*, 372–405. [CrossRef]
44. Peskin, C.S. Flow Patterns Around Heart Valves: A Numerical Method. *J. Comput. Phys.* **1972**, *10*, 252–271. [CrossRef]
45. Gilmanov, A.; Sotiropoulos, F. A hybrid Cartesian/immersed boundary method for simulating flows with 3D, geometrically complex, moving bodies. *J. Comput. Phys.* **2005**, *207*, 457. [CrossRef]
46. Borazjani, I.; Ge, L.; Sotiropoulos, F. Curvilinear immersed boundary method for simulating fluid structure interaction with complex 3D rigid bodies. *J. Comput. Phys.* **2008**, *227*, 7587–7620. [CrossRef] [PubMed]
47. Ge, L.; Sotiropoulos, F. A numerical method for solving the 3D unsteady incompressible Navier–Stokes equations in curvilinear domains with complex immersed boundaries. *J. Comput. Phys.* **2007**, *225*, 1782–1809. [CrossRef] [PubMed]

48. Balay, S.; Buschelman, K.; Gropp, W.D.; Kaushik, D.; Knepley, M.G.; McInnes, L.C.; Smith, B.F.; Zhang, H. PETSc Web Page 2001. Available online: http://www.mcs.anl.gov/petsc (accessed on 16 April 2020).
49. Borazjani, I. Fluid-structure interaction, immersed boundary-finite element method simulations of bio-prosthetic heart valves. *Comput. Methods Appl. Mech. Eng.* **2013**, *257*, 103–116. [CrossRef]
50. Borazjani, I.; Ge, L.; Le, T.; Sotiropoulos, F. A parallel overset-curvilinear-immersed boundary framework for simulating complex 3D incompressible flows. *Comput. Fluids* **2013**, *77*, 76–96. [CrossRef] [PubMed]
51. Asgharzadeh, H.; Borazjani, I. A Newton–Krylov method with an approximate analytical Jacobian for implicit solution of Navier–Stokes equations on staggered overset-curvilinear grids with immersed boundaries. *J. Comput. Phys.* **2017**, *331*, 227–256. [CrossRef]
52. Behara, S.; Borazjani, I.; Sotiropoulos, F. Vortex-induced vibrations of an elastically mounted sphere with three degrees of freedom at Re=300: Hysteresis and vortex shedding modes. *J. Fluid Mech.* **2011**, *686*, 426–450. [CrossRef]
53. Borazjani, I.; Sotiropoulos, F. Vortex-induced vibrations of two cylinders in tandem arrangement in the proximity–wake interference region. *J. Fluid Mech.* **2009**, *621*, 321. [CrossRef]
54. Borazjani, I. The functional role of caudal and anal/dorsal fins during the C-start of a bluegill sunfish. *J. Exp. Biol.* **2013**, *216*, 1658–1669. [CrossRef]
55. Borazjani, I.; Daghooghi, M. The fish tail motion forms an attached leading edge vortex. *Proc. R. Soc. B* **2013**, *280*, 20122071. [CrossRef]
56. Borazjani, I.; Sotiropoulos, F. Numerical investigation of the hydrodynamics of carangiform swimming in the transitional and inertial flow regimes. *J. Exp. Biol.* **2008**, *211*, 1541–1558. [CrossRef]
57. Borazjani, I.; Sotiropoulos, F. Numerical investigation of the hydrodynamics of anguilliform swimming in the transitional and inertial flow regimes. *J. Exp. Biol.* **2009**, *212*, 576–592. [CrossRef] [PubMed]
58. Borazjani, I.; Sotiropoulos, F. Why don't mackerels swim like eels? The role of form and kinematics on the hydrodynamics of undulatory swimming. *Phys. Fluids* **2009**, *21*, 091109. [CrossRef]
59. Borazjani, I.; Sotiropoulos, F. On the role of form and kinematics on the hydrodynamics of self-propelled body/caudal fin swimming. *J. Exp. Biol.* **2010**, *213*, 89–107. [CrossRef] [PubMed]
60. Borazjani, I.; Sotiropoulos, F.; Tytell, E.D.; Lauder, G.V. Hydrodynamics of the bluegill sunfish c-start escape response: Three-dimensional simulations and comparison with experimental data. *J. Exp. Biol.* **2012**, *215*, 671–684. [CrossRef] [PubMed]
61. Daghooghi, M.; Borazjani, I. The hydrodynamic advantages of synchronized swimming in a rectangular pattern. *Bioinspir. Biomim.* **2015**, *10*, 056018. [CrossRef]
62. Daghooghi, M.; Borazjani, I. Self-propelled swimming simulations of bio-inspired smart structures. *Bioinspir. Biomim.* **2016**, *11*, 056001. [CrossRef]
63. Tytell, E.D.; Borazjani, I.; Sotiropoulos, F.; Baker, T.V.; Anderson, E.J.; Lauder, G.V. Disentangling the functional roles of morphology and motion in the swimming of fish. *Integr. Compar. Biol.* **2010**, *50*, 1140–1154. [CrossRef]
64. Borazjani, I.; Ge, L.; Sotiropoulos, F. High-Resolution Fluid–Structure Interaction Simulations of Flow Through a Bi-Leaflet Mechanical Heart Valve in an Anatomic Aorta. *Ann. Biomed. Eng.* **2010**, *38*, 326–344. [CrossRef]
65. Borazjani, I.; Sotiropoulos, F. The effect of implantation orientation of a bi-leaflet mechanical heart valve on kinematics and hemodynamics in an anatomic aorta. *ASME J. Biomech. Eng.* **2010**, *132*, 111005–111008. [CrossRef]
66. Borazjani, I.; Westerdale, J.; McMahon, E.; Rajaraman, P.K.; Heys, J.; Belohlavek, M. Left Ventricular Flow Analysis: Recent Advances in Numerical Methods and Applications in Cardiac Ultrasound. *Comput. Math. Methods Med. Special Issue Comput. Anal. Coronary Ventric. Hemodyn.* **2013**, *2013*, 395081-11. [CrossRef]
67. Le, T.B.; Borazjani, I.; Sotiropoulos, F. Pulsatile Flow Effects on the Hemodynamics of Intracranial Aneurysms. *J. Biomech. Eng.* **2010**, *132*, 111009. [CrossRef] [PubMed]
68. Song, Z.; Borazjani, I. The role of shape and heart rate on the performance of the left ventricle. *J. Biomech. Eng.* **2015**, *137*, 114501. [CrossRef] [PubMed]
69. Hedayat, M.; Borazjani, I. Comparison of platelet activation through hinge vs bulk flow in bileaflet mechanical heart valves. *J. Biomech.* **2019**, *83*, 280–290. [CrossRef] [PubMed]
70. Hedayat, M.; Asgharzadeh, H.; Borazjani, I. Platelet activation of mechanical versus bioprosthetic heart valves during systole. *J. Biomech.* **2017**, *56*, 111–116. [CrossRef] [PubMed]

71. Khosronejad, A.; Kang, S.; Borazjani, I.; Sotiropoulos, F. Curvilinear immersed boundary method for simulating coupled flow and bed morphodynamic interactions due to sediment transport phenomena. *Adv. Water Resour.* **2011**, *34*, 829–843. [CrossRef]
72. Akbarzadeh, A.; Borazjani, I. A numerical study on controlling flow separation via surface morphing in the form of backward traveling waves. In Proceedings of the AIAA Aviation 2019 Forum, Dallas, TX, USA, 17–21 June 2019.
73. Akbarzadeh, A.M.; Borazjani, I. Reducing flow separation of an inclined plate via travelling waves. *J. Fluid Mech.* **2019**, *880*, 831–863. [CrossRef]
74. Akbarzadeh, A.M.; Borazjani, I. Large eddy simulations of a turbulent channel flow with a deforming wall undergoing high steepness traveling waves. *Phys. Fluids* **2019**, *31*, 125107. [CrossRef]
75. Asadi, H.; Asgharzadeh, H.; Borazjani, I. On the scaling of propagation of periodically generated vortex rings. *J. Fluid Mech.* **2018**, *853*, 150–170. [CrossRef]
76. Daghooghi, M.; Borazjani, I. The influence of inertia on the rheology of a periodic suspension of neutrally buoyant rigid ellipsoids. *J. Fluid Mech.* **2015**, *781*, 506–549. [CrossRef]
77. Daghooghi, M.; Borazjani, I. The effects of irregular shape on the particle stress of dilute suspensions. *J. Fluid Mech.* **2018**, *839*, 663–692. [CrossRef]
78. Ardeshiri, H.; Schmitt, F.; Souissi, S.; Toschi, F.; Calzavarini, E. Copepods encounter rates from a model of escape jump behaviour in turbulence. *J. Plankton Res.* **2017**, *39*, 878–890. [CrossRef]
79. Jiang, H.; Meneveau, C.; Osborn, T.R. Numerical study of the feeding current around a copepod. *J. Plankton Res.* **1999**, *21*, 1391. [CrossRef]
80. Jiang, H.; Meneveau, C.; Osborn, T.R. The flow field around a freely swimming copepod in steady motion. Part II: Numerical simulation. *J. Plankton Res.* **2002**, *24*, 191. [CrossRef]
81. Jiang, H.; Osborn, T.R.; Meneveau, C. Hydrodynamic interaction between two copepods: A numerical study. *J. Plankton Res.* **2002**, *24*, 235. [CrossRef]
82. Hunt, J.C.; Wray, A.A.; Moin, P. *Eddies, Streams, and Convergence Zones in Turbulent Flows*; Center for Turbulence Research Report CTR-S88; NASA: Washington, DC, USA, 1988.
83. Bottom II, R.; Borazjani, I.; Blevins, E.; Lauder, G. Hydrodynamics of swimming in stingrays: Numerical simulations and the role of the leading-edge vortex. *J. Fluid Mech.* **2016**, *788*, 407–443. [CrossRef]
84. Jumars, P.A.; Trowbridge, J.H.; Boss, E.; Karp-Boss, L. Turbulence-plankton interactions: A new cartoon. *Mar. Ecol.* **2009**, *30*, 133–150. [CrossRef]

© 2020 by the author. Licensee MDPI, Basel, Switzerland. This article is an open access article distributed under the terms and conditions of the Creative Commons Attribution (CC BY) license (http://creativecommons.org/licenses/by/4.0/).

Article

Nutrient Patchiness, Phytoplankton Surge-Uptake, and Turbulent History: A Theoretical Approach and Its Experimental Validation

Mathilde Schapira [1,*,†] **and Laurent Seuront** [2,3,4,*,†]

[1] Ifremer, LITTORAL, F-44311 Nantes, France
[2] CNRS, Univ. Lille, Univ. Littoral Côte d'Opale, UMR 8187, LOG, Laboratoire d'Océanologie et de Géosciences, F-62930 Wimereux, France
[3] Department of Marine Resource and Energy, Tokyo University of Marine Science and Technology, 4-5-7 Konan, Minato-ku, Tokyo 108-8477, Japan
[4] Department of Zoology and Entomology, Rhodes University, Grahamstown 6140, South Africa
* Correspondence: mathilde.schapira@ifremer.fr (M.S.); laurent.seuront@cnrs.fr (L.S.)
† These authors contributed equally to this work.

Received: 30 March 2020; Accepted: 11 May 2020; Published: 22 May 2020

Abstract: Despite ample evidence of micro- and small-scale (i.e., millimeter- to meter-scale) phytoplankton and zooplankton patchiness in the ocean, direct observations of nutrient distributions and the ecological importance of this phenomenon are still relatively scarce. In this context, we first describe a simple procedure to continuously sample nutrients in surface waters, and subsequently provide evidence of the existence of microscale distribution of ammonium in the ocean. We further show that ammonium is never homogeneously distributed, even under very high conditions of turbulence. Instead, turbulence intensity appears to control nutrient patchiness, with a more homogeneous or a more heterogeneous distribution observed under high and low turbulence intensities, respectively, under the same concentration in nutrient. Based on a modelling procedure taking into account the stochastic properties of intermittent nutrient distributions and observations carried out on natural phytoplankton communities, we introduce and verify the hypothesis that under nutrient limitation, the *"turbulent history"* of phytoplankton cells, i.e., the turbulent conditions they experienced in their natural environments, conditions their efficiency to uptake ephemeral inorganic nitrogen patches of different concentrations. Specifically, phytoplankton cells exposed to high turbulence intensities (i.e., more homogeneous nutrient distribution) were more efficient to uptake high concentration nitrogen pulses (2 µM). In contrast, under low turbulence conditions (i.e., more heterogeneous nutrient distribution), uptake rates were higher for low concentration nitrogen pulses (0.5 µM). These results suggest that under nutrient limitation, natural phytoplankton populations respond to high turbulence intensities through a decrease in affinity for nutrients and an increase in their transport rate, and vice versa.

Keywords: nutrient patchiness; turbulence; phytoplankton; surge uptake; nutrient depletion; turbulent history

1. Introduction

Investigations of micro- to small-scale (typically millimeter- to meter-scale) distributions of viruses, bacteria, phytoplankton, and zooplankton populations revealed their patchy character, see e.g., [1–22], in particular in relation with turbulence [23–25]. Despite early attempts to quantify nutrient patchiness in the ocean [26,27], the introduction of a simple procedure to continuously sample nutrients from surface waters nearly two decades ago [28], and the plethora of work devoted to the assessment of the implication of microscale nutrient patchiness for aquatic microbial communities [29–35], direct observations of micro-scale nutrient distributions are still lacking. Information on the qualitative and quantitative nature of micro-scale nutrient distribution is nevertheless critically needed to bridge the gap between bacteria and phytoplankton and higher trophic levels to improve our general understanding of structures and functions in marine systems.

The existence of nutrient microzones has long been hypothesized to result in the development of phytoplankton adaptive strategies for nutrient uptake [36–38]. However, little is still known about the potential effect of the interplay between nutrient patchiness and phytoplankton uptake in natural waters. The uptake of nutrient by phytoplankton cells is typically described using the Monod equation, i.e., a strict equivalent to the Michaelis-Menten equation. These models both fundamentally hypothesized steady state conditions, i.e., a homogenous distribution of the limiting nutrient in time. Under non-steady state conditions, such as nutrient patchiness, these equations cannot represent correctly nutrient removal by phytoplankton cells and to the best of our knowledge there is no experimentally validated model of nutrient uptake under fluctuating nutrient conditions [39]. Nutrient patchiness may have a negative impact on nutrient uptake rates as uptake is typically less efficient at higher nutrient concentration than at low ones [36]. This hypothesis holds true under the general assumption that the parameters of the Michaelis-Menten kinetics remain constant irrespective of ambient nutrient concentration [39]. It becomes, however, fundamentally unrealistic given the well-known abilities of nutritionally limited phytoplankton cells to enhance their uptake of nutrients in the presence of ephemeral point source [40–42].

In this context, the first aim of this work is to briefly rehearse the description of a simple technique allowing the continuous sampling of nutrient from surface waters [28], to critically assess its potential limitations, and to illustrate its validity to characterize nutrient patchiness in the specific framework of sampling experiments conducted in the eastern English Channel. Our second aim was thus to introduce a modelling procedure that might account for the observed surge-uptake of nutrient based (i) on the detailed stochastic properties of intermittent nutrient distributions, and (ii) on a simple adaptive representation of phytoplankton surge-uptake for nutrients. Finally, we validate our mechanistic hypotheses through a specifically designed field experiment devoted to assess the surge uptake rates of natural phytoplankton communities under ammonium limitations when exposed to ammonium pulses of low and high concentrations.

2. Assessing Nutrient Patchiness in the Ocean

2.1. High-Frequency Nutrient Sampling

In order to continuously investigate the small-scale distribution of ammonium, a series of three sampling experiments have been conducted adrift in the coastal waters of the eastern English Channel in the summers of 1996, 1997, and 1998. Water was continuously taken from a depth of 0.25 m through a seawater intake mounted on a suspended hose located 1 m away from the hull of the vessel, and directly processed in a Technicon Autoanalyzer II [43] by means of a rail wheel pump connected to 1.5-mm diameter plastic tubing with an approximate output of 0.80 mL min^{-1}. The temporal resolution (i.e., 3 s) was chosen as the minimum time interval allowed by the Technicon Autoanalyzer II between two ammonium quantity determinations. Despite early suggestions of its feasibility [43], the approach described in the present work has, to our knowledge, only been used at a lower temporal resolution (1 min) and a smoothing of the output signal associated with the dimension of the pumping apparatus [26], and for the determination of nitrites micro-scale distribution [28]. Data were directly recorded on a PC by means of a data logger system interfaced with the Technicon Autoanalyzer II. Between each time series, the whole plastic tubing was rinsed with HCl 10%, Milli-Q water and the Technicon Autoanalyzer II was calibrated using a standardized nitrogen solution. We recorded 11, 15, and 8 time series of ammonium concentrations of approximately 1-h duration at a sampling frequency of 0.33 Hz in 1996, 1997, and 1998, respectively. Sampled time series show the very intermittent character of ammonium distribution (i.e., a distribution characterized by a few dense patches and a wide range of low-density patches; [28]), whatever the year and the hydrodynamic conditions (Figure 1).

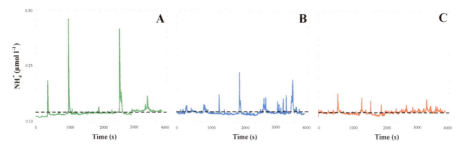

Figure 1. Samples of high-resolution (0.33 Hz) ammonium concentration time series recorded in the eastern English Channel in 1998 for increasing values of the tidal current speed v (m s^{-1}): $v \sim 0.2$ m s^{-1} ($\varepsilon \sim 6 \times 10^{-4}$ m^2 s^{-3}, (**A**)), $v \sim 0.5$ m s^{-1} (i.e., $\varepsilon \sim 7.5 \times 10^{-5}$ m^2 s^{-3}, (**B**)), and $v \sim 1$ m s^{-1} ($\varepsilon \sim 4.8 \times 10^{-6}$ m^2 s^{-3}, (**C**)). The ammonium distributions are patchier when the hydrodynamic conditions (i.e., the tidal current speed v) are weaker. The dissipation rate of turbulent energy ε (m^2 s^{-3}) were estimated from the velocity of the tidal flow v as $\varepsilon = 0.006 \, (v^3/z)$, where z is the depth of the water column ($z \sim 10$ m); see Equation (15) below. The black dashed lines are the average of each distribution, and illustrate the extreme similarity in the bulk concentration of ammonium.

To assess the presence of a potential link between ammonium and phytoplankton biomass that may bias further analysis and interpretation of the distribution of ammonium, the 11, 15, and 8 times series of ammonium concentrations sampled in 1996, 1997, and 1998 were consistently sampled simultaneously to in vivo fluorescence (a proxy of phytoplankton biomass) using a Sea Tech fluorometer, and both temperature and salinity using a Sea-Bird Sealogger CTD. Though the differences in the level of small-scale patchiness observed in nutrients, in vivo fluorescence and purely passive tracers such as temperature and salinity have been discussed at length elsewhere [4,11], we briefly rehearse (i) that we never found any significant correlation ($p > 0.05$) between ammonium times series, and neither in vivo fluorescence, temperature, nor salinity, and (ii) that the stochastic properties of temperature and salinity were consistent with the signature of purely passive tracers advected by turbulent velocity fluctuations, in contrast to both in vivo fluorescence and nutrients that exhibit very distinct levels of patchiness, i.e., more and less patchy under low and high turbulent conditions, respectively [11,28]. These observations warrant that, though ammonium is, *a priori*, a far less conservative nutrient than, e.g., nitrate and nitrite [28], it can then be considered as independent from the phytoplankton and the turbulent fields at the temporal scales considered in the present work.

2.2. Potential Sources of Aliasing and Validation

In order to get a reference framework to validate our continuous use of the Technicon Autoanalyzer II, we plotted the mean of each time series together with ammonium concentrations estimated from triplicated sub-surface Niskin bottle samples taken within 10 cm from the continuous seawater intake at the beginning, the middle and the end of each time series record (Figure 2). Both the very good agreement observed between the ammonium concentration obtained from our novel continuous sampling and a standard discrete sampling scheme and the lack of interannual variability in the observed agreement suggest that the occurrence of the observed high-density patches (cf. Figure 1) cannot be attributed to any kind of aliasing. Several potential sources of aliasing that can be proposed for the occurrence of these patches are nevertheless discussed hereafter.

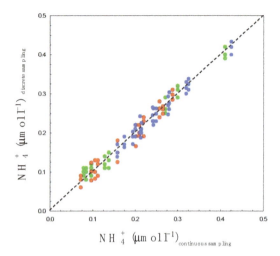

Figure 2. Mean values of ammonium times series recorded in 1996, 1997, and 1998, plotted against the ammonium concentrations simultaneously estimated from sub-surface Niskin bottle samples. The resulting highly significant linear regression ($p < 0.01$) demonstrates the validity of our high-resolution sampling procedure. The different colors correspond to sampling under different flow speed v conditions: red ($v \sim 1$ m s^{-1}; i.e., $\varepsilon \sim 6 \times 10^{-4}$ m^2 s^{-3}), blue ($v \sim 0.5$ m s^{-1}, i.e., $\varepsilon \sim 7.5 \times 10^{-5}$ m^2 s^{-3}), and green ($v \sim 0.2$ m s^{-1}, i.e., $\varepsilon \sim 4.8 \times 10^{-6}$ m^2 s^{-3}). The dissipation rate of turbulent energy ε (m^2 s^{-3}) were estimated from the velocity of the tidal flow v as $\varepsilon = \phi v^3/z$, where z is the depth of the water column ($z \sim 10$ m) and ϕ is a constant, $\phi = 0.006$; see Equation (15) below.

As briefly investigated by Seuront et al. [28], the four main sources of aliasing that can be identified in our sampling procedure are discussed hereafter.

2.2.1. The Motion of the Ship

A contamination by the motion of the ship implies that each point of a recorded dataset may potentially have been sampled at a different depth. Such a bias would typically be related to the characteristic frequency of the waves, and consequently would have led to peaks at specific frequencies when plotting the ammonium time series in Fourier space [44], which is obviously not the case (Figure 3). Note, however, that even if the observed ammonium patches could have been related to the rolling of the ship, their ecological relevance would be equivalent, providing evidence for vertical small-scale patchiness instead of/and a horizontal one.

2.2.2. The Characteristics of the Sample Processing Chain Including Features of the Electronics Involved

In the present case, the linearity of the ammonium power spectra over the whole range of available scales (Figure 3) shows the absence of any kind of noise contaminations by the electronics or the processing chain, in which case the high-frequency part of the spectra would have shown a roll-off towards the noise level of the involved electronics [44]. The absence of this characteristic signature of high-frequency noise demonstrates that our sampling frequency is well above the electronic noise level. On the other hand, both low- and high-density patches perceptible from Figure 1 include more than one point (i.e., from 5 to 20), and this excludes the occurrence of electronic spikiness. These patches correspond to 5 to 20 points, then to time scales bounded between 15 and 60 s, and considering the mean flows experienced during our sampling experiments, to spatial scales bounded between 0.2 and 60 m.

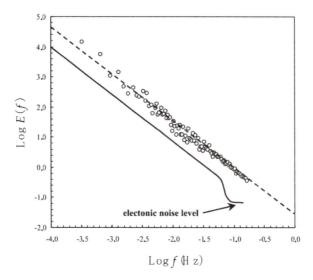

Figure 3. The power spectrum $E(f)$, where f is frequency, of an ammonium time series recorded in 1997. The strong linearity of the power spectrum indicates a scaling behavior over the whole range of scales. The spectrum expected in case of noise contamination by the electronics of the processing chain, presenting a high-frequency roll-off towards the electronic noise level, is shown for comparison.

2.2.3. The Turbulent and/or Molecular Diffusion Occurring in the Plastic Tubing of the Pumping Apparatus

First, we investigated the importance of the forces due to viscosity by calculating the Reynolds number Re, Re = ud/v, where u is the flow speed in the plastic tube (u = 12.5 mm s^{-1}), d the diameter of the tube (d = 1.5 mm), and v is the kinematic viscosity (v = 10^{-6} m^2 s^{-1}). The low Reynolds number associated with our pumping apparatus (Re = 18.75) indicates that no turbulent mixing occurs during the pumping process in the plastic tubing. We subsequently assessed the diffusion length scale, L, travelled by NH_4^+ molecules in the plastic tubing during the pumping process due to the molecular diffusion; L is defined as $L = (Dt)^{0.5}$, where D is the molecular diffusion (D = 10^{-9} m^2 s^{-1}) and t the diffusion time scale (i.e., the time needed to bring seawater through the Technicon Autoanalyzer II; i.e., t = 20, 22, and 16 min for the sampling experiments conducted in 1996, 1997, and 1998, respectively). The diffusion length scale L is about 10^{-3} m (i.e., L = 1.10 × 10^{-3} m in 1996, L = 1.15 × 10^{-3} m in 1997, and L = 0.98 × 10^{-3} m in 1998). This length scale is typically five orders of magnitude lower than the inter-sample bubbling used by the Technicon Autoanalyzer II [28]. As a consequence, it is stressed that the low Reynolds number and minute diffusion length scale characteristic of our pumping procedure cannot induce any bias related to turbulent diffusion nor molecular diffusion.

2.2.4. The Mixing Induced by the Boundary Layer Occurring around the Hull of the Vessel

This potential bias has been investigated by estimating the thickness of the boundary layer generated by the tidal current flowing around the hull of the ship, which have been compared to the 1-m distance chosen for the seawater intake. The thickness of a boundary layer, δ, increases with increasing distance from the ship bow according to $\delta = (xv/v)^{1/2}$ [45], where x is the distance from the ship bow where water has been continuously taken (x = 15 m), v the kinematic viscosity and v (m s^{-1}) the tidal current speed. We then estimated δ for the range of velocities (0.05–1.50 m s^{-1}) experienced during the seawater pumping experiments as being in the range 0.32–1.73 cm. The potential influence of such minute boundary layer thickness on the temporal patterns of the nitrite measurements taken 1 m away can be obviously neglected.

2.3. Stochastic Quantification of Intermittent Nutrient Distribution

2.3.1. Theoretical Analysis

The fluctuations of passive scalars such as temperature, salinity, and a priori phytoplankton cells occurring under the influence of fully developed homogeneous tri-dimenstional turbulence have widely been described in Fourier space using power spectral analysis as $E(f) \propto f^{-\beta}$ where f is the frequency (s^{-1}) and $\beta \approx 5/3$. A power spectrum fundamentally quantifies the amount of variability occurring in different frequency bands. When all or parts of the spectrum follow the abovementioned power law, or more generally a power law of the form $E(f) \propto f^{-\beta}$ where β can diverge from the theoretical expectation $\beta \approx 5/3$, this indicates the absence of any characteristic time scale in the range of scales where the power law applies.

Spectral analysis is, however, limited to a second-order statistic, hence characterizes very poorly intermittent fluctuations (i.e., occasional and unpredictable large peaks separated by very low values; see Figure 1) that are fundamentally non-Gaussian [4]. Spectral analysis can be generalized in real space using the qth order structure functions that have been extensively described and illustrated elsewhere [4]. For a concentration of nutrient C fluctuating in time, the qth order structure functions are defined as $\langle (\Delta C(\tau))^q \rangle \propto \langle |C(t+\tau) - C(\tau)|^q \rangle$, where the quantity $\langle (\Delta C(\tau))^q \rangle$ is the qth order statistical moments of the fluctuations of the quantity C at the time scale τ.

The structure function exponents $\zeta(q)$ are defined as $\langle (\Delta C(\tau))^q \rangle \propto t^{\zeta(q)}$, and the scaling exponents $\zeta(q)$ are given by the slope of the linear trends of $\langle (\Delta C(\tau))^q \rangle$ versus τ in a log–log plot. Specifically, the exponent $\zeta(1) = H$ defines the scale-dependency of the average fluctuations, that is if $H \neq 0$, the fluctuations will depend on the time scale. In turn, the power spectral slope β is linked to the second moment $\zeta(2)$ as $\beta = 1 + \zeta(2)$. The function $\zeta(q)$ is linear for monoscaling (i.e., monofractal) processes such as Brownian motion (($\zeta(q) = q/2$) and non-intermittent turbulence ($\zeta(q) = q/3$). In contrast, $\zeta(q)$ is non-linear and convex for multiscaling (i.e., multifractal) processes [4]. The convexity of the function $\zeta(q)$, i.e., $\zeta(q) = qH - K(q)$, corresponds to the intermittent (i.e., patchy) deviation from homogeneity, in which case $\zeta(q) = qH$. The parameter $K(2)$, often referred to as μ, is called the *intermittency parameter* and is typically in the range 0.1–0.3 for passive scalars in turbulent flows [4,28,46–48].

Practically, a function $\zeta(q)$ diverging from linearity characterizes a heterogeneous distribution with a few dense patches over a wide range of low-density patches. As such the function $\zeta(q)$ is used in this study as an index of nutrient patchiness: the more convex $\zeta(q)$ is, the more patchy or intermittent the nutrient distribution is.

2.3.2. Intermittent Ammonium Distribution vs. Turbulence Intensity

All the time series investigated significantly diverged from a non-intermittent distribution (Figure 4). Their stochastic properties were related to turbulent mixing intensities, with a clear increase in nonlinearity (i.e., an indication of more intermittent distributions) under conditions of decreasing turbulence. The intermittency parameters $K(2)$ consequently significantly differed according to the velocity of the tidal flow v, with $K(2) = 0.21 \pm 0.01$, $K(2) = 0.14 \pm 0.01$, and $K(2) = 0.06 \pm 0.01$ for $v = 0.20$, 0.5, and 1 m s^{-1}, respectively.

These observations confirmed previous observations conducted on nitrite distribution [28] and further showed that (i) NH_4^+ was consistently heterogeneously distributed for turbulence intensities typically ranging from 10^{-7} to 10^{-4} m^2 s^{-3}, and (ii) for similar concentrations NH_4^+ distributions were significantly more heterogeneous for turbulence intensities ranging from 2.1×10^{-7} m^2 s^{-3} to 7.2×10^{-6} m^2 s^{-3} than for turbulence intensities ranging from 4.8×10^{-5} m^2 s^{-3} to 3.4×10^{-4} m^2 s^{-3}.

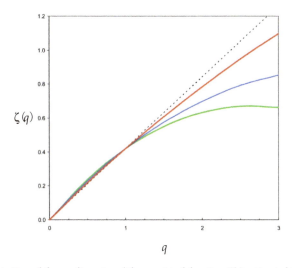

Figure 4. Illustration of the nonlinearity of the empirical function $\zeta(q)$ estimated for ammonium time series under different conditions of flows, i.e., 1 m s^{-1} ($\varepsilon \sim 6 \times 10^{-4}$ m^2 s^{-3}, red), 0.5 m s^{-1} ($\varepsilon \sim 7.5 \times 10^{-5}$ m^2 s^{-3}, blue), and 0.2 m s^{-1} ($\varepsilon \sim 4.8 \times 10^{-6}$ m^2 s^{-3}, green) in comparison to the theoretical linear non-intermittent case $\zeta(q) = qH$ (black dotted line). Note that the empirical function $\zeta(q)$ are consistently non-linear and convex, with an increasing divergence to the theoretical linear case $\zeta(q) = qH$, with increasing values of the statistical order of moment q. The dissipation rate of turbulent energy ε (m^2 s^{-3}) were estimated from the velocity of the tidal flow v as $\varepsilon = \phi v^3/z$, where z is the depth of the water column ($z \sim 10$ m) and ϕ is a constant, $\phi = 0.006$; see Equation (15) below.

3. Nutrient Patches and Phytoplankton Uptake

3.1. Phytoplankton Nutrient Uptake in a Steady-State Environment

The most common functional formulation chosen to describe nutrient uptake rate (J) is the rectangular hyperbola [49]:

$$J = \frac{aC}{b+C} \quad (1)$$

where C is the average extracellular nutrient concentration (mol L^{-1}), a the maximum uptake rate (mol cell^{-1} s^{-1}), and b the extracellular nutrient concentration at half the maximum uptake rate (mol L^{-1}). Equation (1) is strictly equivalent to the Mickaelis-Menten equations used to describe enzyme kinetics mechanistically as:

$$J = \frac{V_{max}C}{K_m + C} \quad (2)$$

where V_{max} is the maximum uptake rate (mol cell^{-1} s^{-1}) and K_m the extracellular nutrient concentration at half the maximum uptake rate (mol L^{-1}). A variety of formulations have further been proposed to take into account the effect of extracellular physical processes, such as fluid motion, molecular diffusivity of nutrient and phytoplankton shape [50,51], and intracellular conditions, such as internal nutrient quotas and rates of biochemical reactions [52], on nutrient uptake. This ultimately led to reformulate nutrient uptake as [53]:

$$J_{i,j} = \psi_j D_i Sh_{i,j} C_i \left(\frac{Q_{i,j}^{max} - Q_{i,j}}{Q_{i,j}^{max}} \right) = \frac{dQ_{i,j}}{dt} \quad (3)$$

where $J_{i,j}$ is the uptake rate of nutrient i to phytoplankton species j (µmol cell^{-1} s^{-1}), ψ_j the diffusion shape factor for phytoplankton species j (m), D_i the molecular diffusivity of chemical species i (m^2 s^{-1}), $Sh_{i,j}$ the Sherwood number (a non-dimensional measure of the passive flux of nutrient i to phytoplankton species j due to local fluid motion), C_i the average extracellular concentration of chemical species i (µmol L^{-1}), $Q_{i,j}^{max}$ the maximum internal cell quota of nutrient i in phytoplankton species j (µmol cell^{-1}), and $Q_{i,j}$ is the internal cell quota of nutrient i in phytoplankton species j (µmol cell^{-1}).

As stressed by Currie [39], and to the best of our knowledge, no experimentally validated model of nutrient uptake under fluctuating nutrient conditions exists. Moreover, Michaelis-Menten kinetics do not satisfactorily explain uptake when nutrient concentrations fluctuate in time. Theoretical works suggested that nutrient patchiness will negatively impact uptake rates as uptake efficiency is lower at high nutrient concentrations than at low ones [39]. This is acceptable under the general assumption that the parameters of the Michaelis-Menten kinetics remain constant irrespective of ambient nutrient concentration [39]. This assumption is, however, unrealistic given the known abilities of nutritionally limited phytoplankton cells to enhance their uptake of nutrients in the presence of ephemeral point source [40–42]. We thus introduce hereafter a novel model that may account for the observed surge-uptake of nutrients based (i) on the detailed stochastic properties of intermittent nutrient distributions, and (ii) on a simple adaptive representation of phytoplankton surge-uptake for nutrients.

3.2. A Simplified Model of Nutrient Surge Uptake in an Intermittent Environment

3.2.1. Theoretical Formulation of the Stochastic Properties of Intermittent Nutrient Distribution

We only review here the main properties of intermittent (i.e., multifractal) fields. More details on the use of multifractal algorithms to marine ecology studies and what can be concluded from their use can be found elsewhere [4,54]. A main property of an intermittent field is that its fluctuations are not destroyed by smoothing at any scale, until the outer scale of the system is reached. For a given nutrient concentration C, this means that the intermittent field C (see Figure 1) averaged over a scale l will have a scale-dependent value denoted as C_l, or as C_λ. Here we introduce a non-dimensional scale ratio λ ($\lambda = L/l$), which is the ratio between an external length scale L and a targeted length scale l within the inertial sub-range, i.e., $l_k \leq l \leq L$ where the Kolmogorov length scale is expresses as l_k; see [4,54] for further details. We assume in this analysis $\lambda \gg 1$. The scale-dependent multifractal field C_λ can be described by its probability distribution, or equivalently, by its statistical moments $\langle (C_\lambda)^q \rangle$, where we consider any $q \geq 0$. These moments can be scaled with the scale ratio λ, as [4,54]:

$$\langle (C_\lambda)^q \rangle = C_0^q \lambda^{K(q)}, \qquad (4)$$

Considering a continuous range of values of $q \geq 0$, Equation (4) is valid only for scales belonging to the inertial subrange, thus for $1 \leq \lambda \leq \Lambda$, where $\Lambda = L/l_B$ is the maximum scale ratio, between the larger outer scale L and the Batchelor scale l_B, i.e., the smallest length scales of fluctuations in scalar concentration (nutrient concentration in the present case) that can exist before being dominated by molecular diffusion. The angle brackets '⟨.⟩' in Equation (4) indicate statistical averaging, $C_0^q = \langle C_\lambda \rangle$ is the mean of the multifractal process C_λ, and $K(q)$ is a concave scale-invariant moment function that satisfies $K(0) = 0$ and $K(1) = 0$ [4]. The function $K(q)$ describes the whole statistics of the process, in an equivalent manner as the probability distribution. As stressed above, the second moment $K(2)$ is usually used as an intermittency parameter and referred to as μ. Subsequently, Equation (4) can be used to evaluate the average of any polynomial function $f(C_\lambda)$ of the multifractal field C_λ as [54]:

$$f(C_\lambda) = \sum_{p=1}^{N} a_p (C_\lambda)^p \qquad (5)$$

where a_p are constants, and p the polynomial order of the function $f(C_\lambda)$ bounded between 1 and N. Averaging the function $f(C_\lambda)$ finally leads to:

$$\langle f(C_\lambda) \rangle = \sum_{p=1}^{N} a_p C_0^p \lambda^{K(p)} \tag{6}$$

3.2.2. A Simplified Model for Nutrient Surge Uptake under Intermittent Conditions

Under the assumption of statistical independence between nutrient, phytoplankton, and turbulent fields [28,54], the uptake of nutrient can be thought as a two-step process: (i) the encounter between a phytoplankton cell and nutrient molecules and (ii) the actual nutrient uptake. By analogy with predator-prey encounter theory, the encounter rate E between a phytoplankton cell and nutrient molecules is expressed as:

$$E = \beta C_\lambda \tag{7}$$

where β is the encounter kernel due to turbulence and behavior and C_λ the ambient intermittent nutrient concentration. Now, once a phytoplankton-nutrient encounter occurred, the ability of phytoplankton cells to enhance their uptake of nutrients in the presence of ephemeral point source was expressed following the general mechanistic formulation of Baird and Emsley [53] as:

$$J \propto E \times C_\lambda \tag{8}$$

where J is the instantaneous nutrient uptake rate (µmol cell^{-1} s^{-1}). Equations (7) and (8) subsequently simply rewrite as:

$$J \propto C_\lambda^2 \tag{9}$$

The average uptake rate of a phytoplankton cell exposed to an intermittent nutrient distribution will further be expressed as:

$$\langle J \rangle_{inter} \propto C_0^2 \lambda^{K(2)} \tag{10}$$

where $C_0 = \langle C_\lambda \rangle$ is the average nutrient concentration experienced by phytoplankton cells. In contrast, the average uptake rate of a phytoplankton cell exposed to a homogenous nutrient distribution is given by:

$$\langle J \rangle_{homo} \propto C_0^2 \tag{11}$$

It directly comes from the comparisons of Equations (10) and (11) that:

$$\langle J \rangle_{inter} > \langle J \rangle_{homo} \tag{12}$$

3.2.3. Surge Uptake under Homogenous and Intermittent Nutrient Distribution: The Turbulent History Hypothesis

We estimated the potential effect of intermittent nutrient distributions on phytoplankton uptake from Equations (10) and (11) as:

$$\frac{\langle J \rangle_{inter}}{\langle J \rangle_{homo}} \propto \lambda^{K(2)}, \tag{13}$$

using the values of the intermittency parameter $K(2)$ and the scale-ratio λ estimated for ammonium distributions under varying conditions of flow velocities. This resulted in increases in the uptake rates by 4.2-fold, 2.65-fold, and 1.48-fold for flow velocities of 0.2, 0.5, and 1 m s^{-1}, respectively. Assuming nutrient limitation, and in the absence of significant differences in the average ammonium concentrations (i.e., $C_0 = \langle C_\lambda \rangle$), these results imply that under elevated turbulent conditions, phytoplankton cells would experience a low-density background of ammonium concentration, resulting in low uptake rates. These cells are then more likely to be nutrient depleted, and would hence exhibit low affinity for NH_4^+ [41,50,55,56]. An optimal uptake strategy would thus consist in an increase in transport rate toward the cell [56,57] through the activation of nitrogen-regulated proteins [58,59]. In contrast, under low turbulent conditions, phytoplankton cells would experience high density nitrogen patches over a wide range of impoverished water, become transport limited in their uptake rates [50,55–57], and develop a higher affinity for NH_4^+. Note that affinity is here employed *sensu* Healey (1980), i.e., low affinity and high affinity respectively relate to high and low values of K_m in Equation (2) [60].

The results of our observations and theoretical model suggest that:

- For the same concentrations, the distribution of ammonium is controlled by turbulence, switching from a more homogeneous to a more heterogeneous distribution respectively under high and low turbulence intensities. This is consistent with previous observations conducted on nitrite and phytoplankton concentrations [11,28].
- The turbulent regime experienced by phytoplankton cells, here referred to as their 'turbulent history' will condition their affinity to ammonium and its transport rate.
- As a consequence, any uptake experiments conducted on natural phytoplankton communities would be intrinsically influenced, if not biased, by their turbulent history. In order to validate our mechanistic hypotheses, we specifically designed a field experiment devoted to assess the surge uptake rates of natural phytoplankton communities under ammonium limitations when exposed to ammonium pulses of low and high concentrations.

4. Empirical Validation: A Case Study from a Turbulent Coastal Sea, the Eastern English Channel

4.1. Field Site and Sampling Strategy

The eastern English Channel (EEC) is a tidally-mixed coastal ecosystem where strong tidal currents and shallow waters lead to turbulent kinetic energy dissipation rates varying between 10^{-6} and 10^{-4} m^2 s^{-3} [11,28]. This area is also structured along a north/south gradient (Figure 5); the northern Strait of Dover being more turbulent than the sheltered waters of the southern Bay of Somme [61].

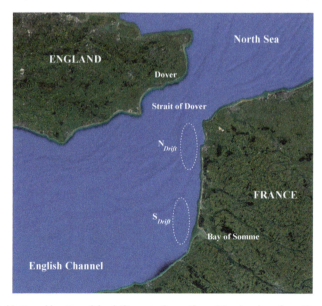

Figure 5. Field site and location of the drift area in the northern (N_{drift}) and southern (S_{drift}) part of the eastern English Channel.

Three multidisciplinary cruises were conducted during the late phase of the 2003 spring phytoplankton bloom (i.e., April, May, and July; Table 1) adrift aboard the N/O "Côtes de la Manche"

(CNRS, INSU) in the Strait of Dover and the Bay of Somme (Table 1, Figure 5). Hereafter, sampling sites are referred to as N and S, where "N" and "S" are the northern (Strait of Dover) and southern (Bay of Somme) water masses of the eastern English Channel. ARGOS buoy equipped with a GPS system was followed during 24 h for each period and location. Temperature (°C) and salinity profiles were acquired at each sampling station with a SBE 25 Sealogger CTD (Sea-Bird Scientific, Bellevue, WA, USA). Water samples were taken from sub-surface (1 m) using 30-L Niskin bottles. No sampling was conducted in the northern station in April (Table 1).

Table 1. In situ conditions during North (N) and South (S) drift cruises; dates, locations (latitude/longitude), depths and tidal conditions (ST: spring tide; NT: neap-tide; F: flood tide and E: ebb tide), vertically averaged (±SE) salinity (S) and temperature (T), nitrite + nitrate ($NO_2^- + NO_3^-$), and ammonium (NH_4^+) concentrations (µM), and chlorophyll a ([Chl a]) surface concentrations (µg L^{-1}), on sampling stations. (-) No data was available for N in April. DL: Detection Limit.

Area	Date	Latitude	Longitude	Depth (m)	Tide	F/E	S(PSU)	T (°C)	Nitrite + Nitrate (mM)	Ammonium (mM)	Chlorophyll a (mg L^{-1})
S	April 24	50°37′688 N	1°25′963 E	17.5	NT	E	33.9 (0.3)	9.9 (0.2)	0.54	0.2	7.22
N	May 11	50°41′052 N	1°26′988 E	33	NT	E	34.7 (0.0)	11.1 (0.1)	<DL	0.4	1.7
S	May 14	50°20′489 N	1°24′633 E	19.5	ST	E	34.1 (0.0)	11.8 (0.0)	0.11	0.78	6.11
N	July 7	50°50′240 N	1°28′192 E	52	NT	E	33.3 (0.2)	18.7 (0.2)	<DL	1	3.94
S	July 9	50°18′432 N	1°22′336 E	15.7	NT	E	34.2 (0.1)	17.6 (0.1)	0.1	0.72	5.11

4.2. Chemical and Biological Environment

For dissolved inorganic nitrogen ($NO_2^- + NO_3^-$), 10-mL water samples were frozen at −20 °C immediately after collection, and analyzed in the laboratory with an auto-analyzer (Alliance Integral Futura, AMS Alliance, Frépillon, France). Ammonium (NH_4^+) concentrations were determined manually on 100-mL water samples [62]. Water samples (200 mL) were filtered through glass-fiber filters (Whatman GF/C, GE Healthcare Life Sciences, Little Chalfront, UK), immediately frozen (−20 °C), and chlorophyllous pigments were subsequently extracted with 90% acetone (5 mL) in the dark at 4 °C, assayed in a spectrophotometer (UVIKON 940, Kontron instruments®, Montigny-le-Bretonneux, Paris, France) and chlorophyll a concentrations calculated following UNESCO standard calculation [63].

4.3. Quantifying Surge Uptake Rates

Ammonium (NH_4^+) was used to characterize surge uptake processes. This nitrogen source was specifically chosen since (i) phytoplankton cells are susceptible to be frequently exposed to pulses particularly under nitrogen limited conditions [64], (ii) various biological processes (i.e., excretion by auto/heterotrophic plankton, bacterial activities) are likely to release NH_4^+ in the water column and (iii) NH_4^+ is the most limiting nitrogen source in the eastern English Channel, see e.g., [65,66]. Sub-surface water was sampled using a 15-L Niskin bottle, and the surge uptake experiment conducted in nine 1-L bottles filled with in situ seawater; three were used as a control (i.e., without enrichment), three received a low ammonium (as $(NH_4)_2SO_4$) concentration pulse (0.5 µM), and the remaining three with a high ammonium concentration pulse (2 µM). All these bottles were subsequently incubated on the boat deck under natural light and temperature. An additional liter was used to assess the standing stock of auto-, hetero-, and mixotrophic protists.

Ammonium concentrations, $[NH_4^+]$, were measured manually [58] on 10-mL sub-samples taken in each incubation bottle before NH_4^+ addition (t_0) and 5 min after the pulse (t_5). Short-term incubations were chosen to observe surge uptake processes, known to take place within a few minutes after a pulse [36,41]. Surge uptake rates ρ_{surge} (µmol(NH_4^+) µgC^{-1} min^{-1}) were estimated from NH_4^+ consumption, and normalized by phytoplankton biomass as:

$$\rho_{surge} = \left[\left(\left[NH_4^+ \right]_{t_5} - \left[NH_4^+ \right]_{t_0} \right) / \Delta t \right] / C \qquad (14)$$

where $\left[NH_4^+ \right]_{t_0}$ and $\left[NH_4^+ \right]_{t_5}$ are NH_4^+ concentrations (µM) at t_0 and t_5, Δt the time of incubation (Δt = 5 min), and C the phytoplankton biomass (µgC L^{-1}) estimated at each sampling site. Phytoplankton biomass C was estimated from the abundance of auto- and hetero/mixotrophic protists. Cells were measured with an eyepiece micrometer and corresponding biovolumes were calculated by relating the shape of organisms to a standard geometric form. Biovolumes were converted to carbon biomass following [67,68].

4.4. Quantifying the Turbulent History of Phytoplankton Cells

The dissipation rate of turbulent energy induced by the tidal flow ε (m^2 s^{-3}) was estimated as [69]:

$$\epsilon = \phi u^3 / z \qquad (15)$$

where ϕ represent the fraction of the tidal energy used for vertical mixing (ϕ = 0.006, [70]), u the drift speed between two successive stations, which corresponds to M2 depth-averaged tidal velocity (m s^{-1}) and z the water column depth (m).

4.5. Turbulent History, Nutrient Patchiness and Phytoplankton Uptake Rates

Salinity showed a stationary behavior fluctuating between 33.3 (N-July) and 34.7 (N-May) over the survey. In contrast, temperature exhibited a clear seasonal cycle, increasing gradually from 9.9 ± 0.2 °C (S-April) to 18.7 ± 0.2 °C (N-July; Table 1). These temperature and salinity values are consistent with previous measurements done at the seasonal scale in the EEC [65,71–74].

Chlorophyll *a* concentrations ranged between 1.70 µg L^{-1} (N-May) and 7.22 µg L^{-1} (S-April; Table 1). These relatively low Chl *a* concentrations are typical of the conditions encountered during the late phase of the spring bloom in the EEC [72,73]. The composition of phytoplankton populations was homogeneous and dominated by the large diatoms *Guinardia striata* and *Rhizosolenia imbricata* over the duration of the survey [61]. In accordance with previous studies conducted in the EEC [61,62], nitrogen concentrations ($NO_2^- + NO_3^-$ and NH_4^+) remained low (i.e., typically ≤ 1 µM) throughout the survey (Table 1), suggesting that phytoplankton communities were potentially nitrogen-limited.

Turbulent energy dissipation rates ε were highly variable over the survey, ranging between 1.26 × 10^{-6} m^2 s^{-3} (S-April) and 8.59 × 10^{-5} m^2 s^{-3} (N-July; Figure 6). These values are in the range of turbulence intensities reported in this area, i.e., 10^{-7} < ε < 10^{-4} m^2 s^{-3} [11,48]. Based on integrated turbulence intensities experienced by phytoplankton cells during the 5–6 h preceding their sampling, we discriminated two groups of stations: (i) N-July, S-May, and S-July characterized by high turbulent levels (i.e., ε > 10^{-5} m^2 s^{-3}) and (ii) N-May and S-April characterized by lower turbulence intensities (i.e., ε < 10^{-5} m^2 s^{-3}; Figure 6). This critical turbulence intensity has been chosen as it has previously been identified as the turbulence threshold above and below which the level of patchiness of nitrite distributions were significantly different [28]. Specifically, these early observations conducted on nitrite time series were confirmed and specified by the ammonium distributions continuously sampled in 1996, 1997, and 1998. These distributions were consistently patchy irrespective of turbulence intensity (see Figures 1 and 4), and for similar (NH_4^+) concentrations, were significantly more heterogeneous for turbulence intensities ranging from 2.1 × 10^{-7} m^2 s^{-3} to 7.2 × 10^{-6} m^2 s^{-3} than for turbulence intensities ranging from 4.8 × 10^{-5} m^2 s^{-3} to 3.4 × 10^{-4} m^2 s^{-3}.

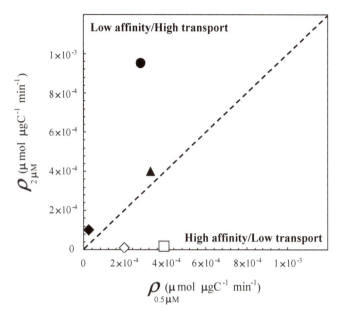

Figure 6. Surge uptake rates after a pulse of 2 µM ($\rho_{2\mu M}$; µmol µgC^{-1} min^{-1}) vs. surge uptake rates after a pulse of 0.5 µM ($\rho_{0.5\mu M}$; µmol µgC^{-1} min^{-1}). The closed symbol are the surge uptake rates measured at stations characterized by high turbulence levels ε (i.e., $\varepsilon > 10^{-5}$ m^2 s^{-3}); S-May (diamond), S-July (triangle) and N-July (circle). The open symbols are the surge uptake rates measured at stations characterized by low turbulence levels (i.e., $\varepsilon < 10^{-5}$ m^2 s^{-3}); N-May (diamond) and S-April (square). The size of the symbol represents the 99% confidence intervals of triplicated surge uptake rates.

The control experiment run without (NH$_4^+$) enrichment did not exhibit any surge uptake, i.e., $\rho_{surge} = 0$; see Equation (14). Surge uptake rates measured following NH$_4^+$ enrichment ranged from 0 to 9.54 × 10^{-4} µmol(NH$_4^+$) µgC^{-1} min^{-1} and appeared to be strongly related to the interplay between NH$_4^+$ concentrations and turbulence intensities (Figure 6). Cells exposed to high turbulent intensities exhibited maximum surge uptake rates after a pulse of 2 µM (Figure 6), suggesting a low affinity for NH$_4^+$ and a subsequent high transport rate. In contrast, cells exposed to lower turbulence intensities before the uptake experiment showed higher surge uptake rates after a pulse of 0.5 µM (Figure 6). This suggests a high affinity for NH$_4^+$ and a low transport rate of NH$_4^+$. Noticeably, no surge uptake was observed following a 2 µM NH$_4^+$ enrichment under the two lowest conditions of turbulence history (Figure 6). While the resolution of this specific issue is beyond the scope of this preliminary study, this observation suggests that turbulence history may play a more critical role than the actual density of the nutrient patches encountered by phytoplankton cells in conditioning their surge uptake abilities. This hypothesis is actually further supported by our observations that showed a consistently higher surge uptake under conditions of high (i.e., 2 µM) NH$_4^+$ enrichment and low turbulent history than under conditions of low (i.e., 0.5 µM) NH$_4^+$ enrichment and high turbulent history.

An undisputed effect of microscale turbulence on phytoplankton is the shear-controlled increase in the passive nutrient fluxes towards phytoplankton cells with turbulence intensities [51,74]. The surge uptake patterns observed here may, however, also rely on the interplay between turbulence and the spatial distribution of dissolved nitrogen. The distribution of dissolved inorganic nutrients has thus been shown to be controlled by the intensity of turbulent mixing, high and low nutrient patchiness were identified under low and high turbulent conditions, respectively [28]; see also Figures 1 and 4.

At scales relevant to individual cells, phytoplankton cells exposed to high and low turbulent intensities may then be adapted to more evenly and more patchy nutrient distributions, respectively.

Under nutrient limitation, phytoplankton cells exposed to high turbulence intensities would experience a low background of evenly distributed nutrients and would thus exhibit low affinity for NH_4^+ [41,50,55,56]. An optimal uptake strategy would thus consist in an increase in transport rate toward the cell [56,57] through the activation of nitrogen regulated proteins, e.g., [58,59]. To date, there is no evidence of the role of microscale shear on the activation of nitrogen regulated proteins, and it is therefore highly speculative, although tempting, to suggest a connection between the elevated shear γ ($\gamma = (\varepsilon/\nu)^{0.5}$, where ε is the dissipation rate of turbulent energy and ν the kinematic viscosity, ca. 10^{-6} m^2 s^{-1}) experienced by phytoplankton cells prior to sampling ($\gamma > 3.2$ s^{-1}, i.e., when $\varepsilon > 10^{-5}$ m^2 s^{-3}) and an increase in transport rates. In contrast, cells exposed to low turbulent intensities would experience high density nitrogen patches over a wide range of impoverished water, become transport limited in their uptake rates [41,50,55,56] and develop a higher affinity for NH_4^+, an assumption consistent with our observations (Figure 6).

5. Conclusions

These results indicate that turbulence controls the microscale distribution of ammonium, switching from a more homogeneous to a more heterogeneous distribution respectively under high and low turbulence intensities for the same bulk concentration. In addition, using a new theoretical framework, we provide evidence of a potential interaction between small-scale turbulence, nutrient patchiness, and nutrient uptake by phytoplankton in natural waters. These preliminary results support the hypothesis that phytoplankton cells exposed to different turbulence levels would exhibit different abilities to use ephemeral nitrogen patches particularly under nitrogen limitation. Turbulence history is thus suggested as a potential fundamental lynch-pin in the control of the nutritive status of phytoplankton cells and as a consequence, any uptake experiments conducted on natural phytoplankton communities would be intrinsically influenced by their turbulent history. More fundamentally, these results highlight the importance of ocean variability at minute spatial and temporal scales in the structure and function of marine ecosystems. In this context, the approach presented here is consistent with previous studies showing how the understanding and subsequent modelling of intermittent distributions—or more generally small- and micro-scale variability—can enhance trophic transfer, interspecific competition, and eventually sustain biodiversity in plankton ecosystems [74–79].

It is finally stressed that simultaneous measurements of small-scale nutrient distributions (see e.g., [28]) and surge uptake rates from different environments, in particular oligotrophic ones, are needed to generalize our observations. The key role played by phytoplankton nutritive status in phytoplankton succession and species success (see e.g., [80]) nevertheless suggests that investigations of the effect of turbulent history on phytoplankton uptake rates may comprise areas of important future research.

Author Contributions: Conceptualization, M.S. and L.S.; methodology, M.S. and L.S.; laboratory analysis, M.S.; theoretical model development, L.S.; data interpretation, M.S. and L.S.; writing—original draft preparation, M.S.; writing—final draft preparation, review, and editing, M.S. and L.S.; funding acquisition, L.S. All authors have read and agreed to the published version of the manuscript.

Funding: This research was funded by the CPER "Phaeocystis", the PNEC "Chantier Manche Orientale-Sud Mer du Nord", and further supported under Australian Research Council's Discovery Projects funding scheme (project numbers DP0664681 and DP0988554). Professor Seuront is the recipient of an Australian Professorial Fellowship (project number DP0988554).

Acknowledgments: We acknowledge the captain and the crew of the NO "Côtes de la Manche" (CNRS-INSU) for their help during the sampling experiment. Two anonymous reviewers are acknowledged for their insightful comments and suggestions that greatly improved the quality of this work.

Conflicts of Interest: The authors declare no conflicts of interest.

References

1. Mitchell, J.G.; Okubo, A.; Fuhrman, J.A. Microzones surrounding phytoplankton forms the basis for a stratified microbial ecosystem. *Nature* **1985**, *316*, 58–59. [CrossRef]
2. Bjørnsen, P.K.; Nielsen, T.G. Decimeter scale heterogeneity in the plankton during a pycnocline bloom of *Gyrodinium aureolum*. *Mar. Ecol. Prog. Ser.* **1991**, *73*, 263–267. [CrossRef]
3. Pascual, M.; Ascioti, A.; Caswell, H. Intermittency in the plankton: A multifractal analysis of zooplankton biomass variability. *J. Plankton Res.* **1995**, *17*, 1209–1232. [CrossRef]
4. Seuront, L.; Schmitt, F.; Shertzer, D.; Lagadeuc, Y.; Lovejoy, S. Universal multifractal analysis as a tool to characterize multiscale intermittent patterns: Example of phytoplankton distribution in turbulent coastal waters. *J. Plankton Res.* **1999**, *21*, 877–922. [CrossRef]
5. Seymour, J.R.; Mitchell, J.G.; Seuront, L. Microscale heterogeneity in the activity of coastal bacterioplankton communities. *Aquat. Microb. Ecol.* **2004**, *35*, 1–16. [CrossRef]
6. Wolk, F.; Yamazaki, H.; Seuront, L. A new free fall profiler for measuring biophysical microstructure. *J. Atmos. Ocean. Technol.* **2002**, *19*, 780–793. [CrossRef]
7. Franks, P.J.S.; Jaffe, J.S. Microscale distributions of phytoplankton: Initial results from a two-dimensional imaging fluorometer. *Mar. Ecol. Prog. Ser.* **2001**, *220*, 59–72. [CrossRef]
8. Seuront, L.; Lagadeuc, Y. Multiscale patchiness of the calanoid copepod *Temora longicornis* in a turbulent coastal sea. *J. Plankton Res.* **2001**, *23*, 1137–1145. [CrossRef]
9. Waters, R.L.; Mitchell, J.G.; Seymour, J.R. Geostatistical characterization of centimeter-scale spatial structure of in vivo fluorescence. *Mar. Ecol. Prog. Ser.* **2003**, *251*, 49–58. [CrossRef]
10. Seymour, J.R.; Mitchell, J.G.; Seuront, L. Microscale and small scale temporal dynamics of a coastal microbial community. *Mar. Ecol. Prog. Ser.* **2005**, *300*, 21–37. [CrossRef]
11. Seuront, L. Hydrodynamic and tidal controls of small-scale phytoplankton patchiness. *Mar. Ecol. Prog. Ser.* **2005**, *302*, 93–101. [CrossRef]
12. Seymour, J.R.; Seuront, L.; Doubell, M.J.; Waters, R.L.; Mitchell, J.G. Microscale patchiness of virioplankton. *J. Biol. Assoc. Ass. UK* **2006**, *86*, 551–561. [CrossRef]
13. Seymour, J.R.; Seuront, L.; Mitchell, J.G. Microscale gradients of planktonic microbial communities above the sediment surface in a mangrove estuary. *Est. Coast. Shelf Sci.* **2007**, *73*, 651–666. [CrossRef]
14. Seymour, J.R.; Seuront, L.; Doubell, M.J.; Mitchell, J.G. Mesoscale and microscale spatial variability of bacteria and viruses during a Phaeocystis globose bloom in the eastern English Channel. *Est. Coast. Shelf Sci.* **2008**, *80*, 589–597. [CrossRef]
15. Doubell, M.J.; Seuront, L.; Seymour, J.R.; Patten, N.L.; Mitchell, J.G. High-resolution fluorometer for mapping microscale phytoplankton distribution. *Appl. Environ. Microbiol.* **2006**, *72*, 4475–4478. [CrossRef] [PubMed]
16. Seuront, L.; Lacheze, C.; Doubell, M.J.; Seymour, J.R.; Dongen-Vogels, V.V.; Newton, K.; Alderkamp, A.C.; Mitchell, J.G. The influence of *Phaeocystis globosa* on microscale spatial patterns of chlorophyll a and bulk-phase seawater viscosity. *Biogeochemistry* **2007**, *83*, 173–188. [CrossRef]
17. Doubell, M.J.; Yamazaki, H.; Hua, L.; Kokubu, Y. An advanced laser-based fluorescence microstructure profiler (TurboMAP-L) for measuring bio-physical coupling in aquatic systems. *J. Plankton Res.* **2009**, *31*, 1441–1452. [CrossRef]
18. Doubell, M.J.; Prairie, J.C.; Yamazaki, H. Millimeter scale profiles of chlorophyll fluorescence: Deciphering the microscale spatial structure of phytoplankton. *Deep-Sea Res. II* **2014**, *101*, 207–215. [CrossRef]
19. Prairie, J.C.; Franks, P.J.S.; Jaffe, J.S. Cryptic peaks: Invisible vertical structure in fluorescent particles revealed using a planar laser imaging fluorometer. *Limnol. Oceanogr.* **2010**, *55*, 1943–1958. [CrossRef]
20. Prairie, J.C.; Franks, P.J.S.; Jaffe, J.S.; Doubell, M.J.; Yamazaki, H. Physical and biological controls of vertical gradients in phytoplankton. *Limnol. Oceanogr. Fluids Environ.* **2011**, *1*, 75–90. [CrossRef]
21. Prairie, J.C.; Sutherland, K.R.; Nickols, K.J.; Kaltenberg, A.M. Biophysical interactions in the plankton: A cross-scale review. *Limnol. Oceanogr. Fluids Environ.* **2012**, *2*, 121–145. [CrossRef]
22. Prairie, J.C.; Ziervogel, K.; Camassa, R.; McLaughlin, R.M.; White, B.L.; Johnson, Z.I.; Arnosti, C. Ephemeral aggregate layers in the water column leave lasting footprints in the carbon cycle. *Limnol. Oceanogr. Lett.* **2017**, *2*, 202–209. [CrossRef]
23. Reigada, R.; Hillary, R.M.; Bees, M.A.; Sancho, J.M.; Sagués, F. Plankton blooms induced by turbulent flows. *Proc. R. Soc. Lond. B* **2003**, *270*, 875–880. [CrossRef] [PubMed]

24. Durham, W.M.; Climent, E.; Barry, M.; De Lillo, F.; Boffetta, G.; Cencini, M.; Stocker, R. Turbulence drives microscale patches of motile phytoplankton. *Nat. Commun.* **2013**, *4*, 2148. [CrossRef] [PubMed]
25. Breier, R.E.; Lalescu, C.C.; Waas, D.; Wilczek, M.; Mazza, M.G. Emergence of phytoplankton patchiness at small scales in mild turbulence. *Proc. Nat. Acad. Sci. USA* **2018**, *115*, 12112–12117. [CrossRef] [PubMed]
26. Estrada, M.; Wagensberg, M. Spatial analysis of spatial series of oceanographic variables. *J. Exp. Mar. Biol. Ecol.* **1977**, *30*, 147–164. [CrossRef]
27. Steele, J.H.; Henderson, E.W. Spatial patterns in North Sea plankton. *Deep-Sea Res.* **1979**, *26*, 955–963. [CrossRef]
28. Seuront, L.; Gentilhomme, V.; Lagadeuc, Y. Small-scale nutrient patches in tidally mixed coastal waters. *Mar. Ecol. Prog. Ser.* **2002**, *232*, 29–44. [CrossRef]
29. Blackburn, N.; Fenchel, T.; Mitchell, J.G. Microscale nutrient patches in planktonic habitats shown by chemotactic bacteria. *Science* **1998**, *282*, 2254–2256. [CrossRef]
30. Stocker, R.; Seymour, J.R.; Samadani, A.; Hunt, D.; Polz, M. Rapid chemotactic response enables marine bacteria to exploit ephemeral microscale nutrient patches. *Proc. Nat. Acad. Sci. USA* **2008**, *105*, 4209–4214. [CrossRef]
31. Seymour, J.R.; Ahmed, T.; Marcos, M.; Stocker, R. A microfluidic chemotaxis assay to study microbial behavior in diffusing nutrient patches *Limnol. Oceanogr. Methods* **2008**, *6*, 477–488. [CrossRef]
32. Seymour, J.R.; Ahmed, T.; Stocker, R. Bacterial chemotaxis towards the extracellular products of the toxic phytoplankton *Heterosigma akashiwo*. *J. Plankton Res.* **2009**, *31*, 1557–1561. [CrossRef]
33. Seymour, J.R.; Marcos, M.; Stocker, R. Resource Patch Formation and Exploitation throughout the Marine Microbial Food Web. *Am. Nat.* **2009**, *173*, E15–E29. [CrossRef]
34. Mitchell, J.G.; Seuront, L.; Doubell, M.J.; Losic, D.; Voelcker, N.H.; Seymour, J.R.; Lal, R. The role of diatom nanostructures in biasing diffusion to improve uptake in a patchy nutrient environment. *PLoS ONE* **2013**, *8*, e59548. [CrossRef]
35. Lambert, B.S.; Raina, J.B.; Fernandez, V.I.; Rinke, C.; Siboni, N.; Rubino, F.; Hugenholtz, P.; Tyson, G.W.; Stocker, R. A microfluidics-based in situ chemotaxis assay to study the behavior of aquatic microbial communities. *Nat. Microbiol.* **2017**, *2*, 1344–1349. [CrossRef]
36. McCarthy, J.J.; Goldman, J.C. Nitrogenous nutrition of Marine phytoplankton in nutrient-depleted waters. *Science* **1979**, *203*, 670–672. [CrossRef]
37. Glibert, P.M.; Goldman, J.C. Rapid ammonium uptake by Marine phytoplankton. *Mar. Biol. Lett.* **1981**, *2*, 25–31.
38. Goldman, J.C.; Glibert, P. Comparative rapid ammonium uptake by four species of marine phytoplankton. *Limnol. Oceanogr.* **1982**, *27*, 814–827. [CrossRef]
39. Currie, D.J. Microscale nutrient patches: Do they matter to the phytoplankton? *Limnol. Oceanogr.* **1984**, *29*, 211–214. [CrossRef]
40. Conway, H.L.; Harrison, P.J.; Davis, C.O. Marine diatoms grown in chemostats under silicate and ammonium limitation. II. Transient response of *Skeletonema costatum* to a single addition of the limiting nutrient. *Mar. Biol.* **1976**, *5*, 187–199. [CrossRef]
41. Collos, Y. Transient situations in nitrate assimilation by marine diatoms. IV. Non-linear phenomena and the estimation of the maximum uptake rate. *J. Plankton Res.* **1983**, *5*, 677–691. [CrossRef]
42. Raimbault, P.; Gentilhomme, V.; Slawyk, G. Short-term responses of 24-hour N-starved cultures of Phaeodactylum tricornutum to pulsed additions of nitrate at nanomolar levels. *Mar. Ecol. Prog. Ser.* **1990**, *63*, 47–52. [CrossRef]
43. Treguer, P.; Le Corre, P. *Manuel D'analyse des sels Nutritifs dans L'eau de mer (Utilisation de L'autoanalyseur II Technicon)*; Université de Bretagne Occidentale: Brest, France, 1971.
44. Jenkins, G.; Watts, D. *Spectral Analysis and its Applications*; Holden-Day: San Francisco, CA, USA, 1968.
45. Mann, K.H.; Lazier, J.R.N. *Dynamic of Marine Ecosystems. Biological-Physical Interactions in the Oceans*; Blackwell: Boston, MA, USA, 1991.
46. Seuront, L.; Schmitt, F.; Shertzer, D.; Lagadeuc, Y.; Lovejoy, S. Multifractal intermittency of Eulerian and Lagrangian turbulence of ocean temperature and plankton fields. *Nonlinear Process. Geophys.* **1996**, *3*, 236–246. [CrossRef]
47. Seuront, L.; Schmitt, F.; Shertzer, D.; Lagadeuc, Y.; Lovejoy, S.; Frontier, S. Multifactal analysis of phytoplankton biomass and temperature in the ocean. *Geophys. Res. Lett.* **1996**, *23*, 3591–3594. [CrossRef]

48. Seuront, L. Space-Time Heterogeneity and Biophysical Coupling in Pelagic Ecology: Implications on Carbon Fluxes. Ph.D. Thesis, Université des Sciences et Techniques de Lille, Villeneuve-d'Ascq, France, 1999.
49. Dugdale, R.C. Nutrient limitation in the sea: Dynamics, identification and significance. *Limnol. Oceanogr.* **1967**, *12*, 685–695. [CrossRef]
50. Pasciak, W.J.; Gavis, J. Transport limited nutrient uptake rates in *Ditylium Brightwellii*. *Limnol. Oceanogr.* **1975**, *20*, 604–617. [CrossRef]
51. Karp-Boss, L.; Boss, E.; Jumars, P.A. Nutrients fluxes to planktonic osmotrophs in the presence of fluid motion. *Oceanogr. Mar. Biol. Ann. Rev.* **1996**, *34*, 71–107.
52. Droop, M.R. Vitamine B-12 and marine ecology IV. The kinetic of uptake, growth and inhibition in *Monochrysis lutheri*. *J. Mar. Biol. Assoc. UK* **1968**, *48*, 689–733. [CrossRef]
53. Baird, M.E.; Emsley, S.M. Toward a mechanistic model of plankton population dynamics. *J. Plankton Res.* **1999**, *21*, 85–126. [CrossRef]
54. Seuront, L.; Schmitt, F.; Lagadeuc, L. Turbulence intermittency, small-scale phytoplankton patchiness and encounter rates in plankton: Where do we go from here? *Deep Sea Res. I* **2001**, *48*, 1199–1215. [CrossRef]
55. Pasciak, W.J.; Gavis, J. Transport limitation of nutrient uptake in phytoplankton. *Limnol. Oceanogr.* **1974**, *19*, 881–888. [CrossRef]
56. Wheeler, P.A.; Glibert, P.M.; McCarthy, J.J. Ammonium uptake incorporation by Chesapeake Bay phytoplankton: Short term uptake kinetics. *Limnol. Oceanogr.* **1982**, *27*, 1113–1128. [CrossRef]
57. Zehr, J.P.; Falkowski, P.G.; Fowler, J.; Capone, D.G. Coupling between ammonium uptake and incorporation in a marine diatom: Experiments with the short-lived radioisotope 13N. *Limnol. Oceanogr.* **1988**, *33*, 518–527. [CrossRef]
58. Antia, N.J.; Harrison, P.J.; Oliveira, L. The role of dissolved organic nitrogen in phytoplankton nutrition, cell biology and ecology. *Phycologia* **1991**, *30*, 1–89. [CrossRef]
59. Palenik, B.; Koke, J.A. Characterisation of a nitrogen-regulated protein identified by cell-surface biotinylation of a marine phytoplankton. *Appl. Environ. Microbiol.* **1995**, *61*, 3311–3315. [CrossRef]
60. Healey, F.P. Slope of the Monod equation as an indicator of advantage in nutrient competition. *Microb. Ecol.* **1980**, *5*, 281–286. [CrossRef]
61. Schapira, M. Space and time dynamic of *Phaeocystis globosa* in the Eastern English Channel: Impact of Turbulence and Sporadic Nutrients Inputs. Ph.D. Thesis, Université des Sciences et Techniques de Lille, Villeneuve-d'Ascq, France, 2005.
62. Koroleff, F. Direct determination of ammonia in natural waters as indophenol blue. *Int. Cons. Explor. Sea* **1969**, *9*, 1–6.
63. UNESCO. Determination of photosynthetic pigments. In *Seawater. -Rep. SCOR/UNESCO Working Group 17. Monographs on Oceanographic Methology*; UNESCO: Paris, France, 1966; pp. 1–69.
64. Raimbault, P.; Gentilhomme, V. Short- and long-term responses of marine diatom *Phaeodactylum tricornutum* to spike addition of nitrate at nanomolar levels. *J. Exp. Mar. Biol. Ecol.* **1990**, *135*, 161–176. [CrossRef]
65. Brunet, C.; Brylinski, J.M.; Frontier, S. Productivity, photosynthetic pigments and hydrology in the coastal front of the Eastern English Channel. *J. Plankton Res.* **1992**, *14*, 1541–1552. [CrossRef]
66. Gentilhomme, V.; Lizon, F. Seasonal cycle of nitrogen and phytoplankton biomass in a well-mixed coastal system (Eastern English Channel). *Hydrobiologia* **1998**, *361*, 191–199. [CrossRef]
67. Menden-Deuer, S.E.; Lessard, E.J. Carbon to volume relationships for dinoflagellates, diatoms, and other protist plankton. *Limnol. Oceanogr.* **2000**, *45*, 569–579. [CrossRef]
68. Menden-Deuer, S.E.; Lessard, J.; Satterberg, J. Effect of preservation on dinoflagellates and diatom cell volume and consequences for carbon biomass predictions. *Mar. Ecol. Prog. Ser.* **2001**, *222*, 41–50. [CrossRef]
69. MacKenzie, B.R.; Legget, W.C. Wind-based models for estimating the dissipation rates of turbulence energy in aquatic environments: Empirical comparisons. *Mar. Ecol. Prog. Ser.* **1993**, *94*, 207–216. [CrossRef]
70. Bowers, D.G.; Simpsons, J.H. Mean position of tidal fronts in European-shelf seas. *Cont. Shelf Res.* **1987**, *7*, 35–44. [CrossRef]
71. Brylinski, J.M.; Brunet, C.; Bentley, D.; Thoumelin, G.; Hilde, D. Hydrography and phytoplankton biomass in the Eastern English Channel in spring 1992. *Est. Coast. Shelf Sci.* **1996**, *43*, 507–519. [CrossRef]
72. Breton, E.; Brunet, C.; Sautour, B.; Brylinski, J.M. Annual variations of phytoplankton biomass in the Eastern English Channel: Comparison by pigment signatures and microscopic counts. *J. Plankton Res.* **2000**, *22*, 1423–1440. [CrossRef]

73. Seuront, L.; Vincent, D.; Mitchell, J.G. Biologically-induced modification of seawater viscosity in the Eastern English Channel during a Phaeocystis globosa spring bloom. *J. Mar. Syst.* **2006**, *61*, 118–133. [CrossRef]
74. Mandal, S.; Locke, C.; Tanaka, M.; Yamazaki, H. Observations and models of highly intermittent phytoplankton distributions. *PLoS ONE* **2014**, *9*, e94797. [CrossRef]
75. Mandal, S.; Homma, H.; Priyadarshi, A.; Burchard, H.; Smith, S.L.; Wiritz, K.W.; Yamazaki, H. A 1D physical-biological model of the impact of highly intermittent phytoplankton distributions. *J. Plankton Res.* **2016**, *38*, 964–976. [CrossRef]
76. Priyadarshi, A.; Mandal, S.; Smith, L.; Yamazaki, H. Micro-scale variability enhances trophic transfer and potentially sustains biodiversity in plankton ecosystems. *J. Theor. Biol.* **2017**, *412*, 86–93. [CrossRef]
77. Smith, S.L.; Mandal, S.; Priyadarshi, A.; Chen, B.; Yamazaki, H. *Modeling the Combined Effects of Physiological Flexibility and Micro-Scale Variability for Plankton Ecosystem Dynamics, Encyclopedia of Ocean Sciences*, 3rd ed.; Cochran, H.K., Bokuniewicz, H.J., Yager, P.L., Eds.; Elsevier: Amsterdam, The Netherlands, 2019; Volume 5, pp. 527–535.
78. Mandal, S.; Lan Smith, S.; Priyadarshi, A.; Yamazaki, H. Micro-scale variability impacts the outcome of competition between different modelled size classes of phytoplankton. *Front. Mar. Sci.* **2019**, *6*, 259. [CrossRef]
79. Priyadarshi, A.; Lan Smith, S.; Mandal, S.; Tanaka, M.; Yamazaki, H. Micro-scale patchiness enhances trophic transfer efficiency and potential plankton biodiversity. *Sci. Rep.* **2019**, *9*, 17243. [CrossRef]
80. Rees, A.P.; Joint, I.; Donald, K.M. Early spring bloom phytoplankton-nutrient dynamics at the Celtic Sea Shelf Edge. *Deep Sea Res. II* **1999**, *46*, 483–510. [CrossRef]

© 2020 by the authors. Licensee MDPI, Basel, Switzerland. This article is an open access article distributed under the terms and conditions of the Creative Commons Attribution (CC BY) license (http://creativecommons.org/licenses/by/4.0/).

Article

Fluid and Predator-Prey Interactions of Scyphomedusae Fed Calanoid Copepods

Zachary Wagner [1], John H. Costello [2,3] and Sean P. Colin [1,3,*]

1. Marine Biology, Roger Williams University, Bristol, RI 02809, USA; zwagner963@g.rwu.edu
2. Biology Department, Providence College, Providence, RI 02918, USA; costello@providence.edu
3. Whitman Center, Marine Biological Laboratory, Woods Hole, MA 02543, USA
* Correspondence: scolin@rwu.edu

Received: 13 March 2020; Accepted: 21 April 2020; Published: 25 April 2020

Abstract: The feeding current of scyphomedusae entrains and transports surrounding fluids and prey through trailing tentacles to initiate encounters with prey. After contact, most prey are retained for ingestion. However, the probability that a contact will occur depends on several factors including capture surface morphology, prey size and behavior. We examined how hydrodynamics, capture surface morphology and prey behavior affect the capture probability of copepods. To do this, we documented medusa-copepod interactions of four species of scyphomedusae (two semeostomes and two rhizostomes) possessing different capture surface morphologies. We tracked the movement and behavior of entrained copepods throughout the feeding process to quantify prey behavior effects upon capture efficiency (# captures/# encounters). The feeding currents generated by all the medusan species generated fluid shear deformation rates well above the detection limits of copepods. Despite strong hydrodynamic signals, copepod behavior was highly variable and only 58% of the copepods reacted to entrainment within feeding currents. Furthermore, copepod behavior (categorized as no reaction, escape jump or adjustment jump) did not significantly affect the capture efficiency. The scale and complexity of the feeding current generated by scyphomedusae may help explain the poor ability of copepods to avoid capture.

Keywords: jellyfish; hydrodynamics; escape behavior; *Acartia tonsa*

---

1. Introduction

Scyphomedusae are influential predators within marine ecosystems, capable of consuming a high abundance and a diverse array of prey items [1,2]. Bell contractions by scyphomedusae create feeding currents that draw surrounding fluid to capture surfaces. Zooplankton within the entrained fluid are drawn into the tentacles or oral arms and, once physically contacted, are unlikely to escape [3].

While physical contact between these predators and their prey usually leads to prey capture, there is no guarantee that entrained prey will actually contact a capture surface. In fact, that probability appears to be relatively low and quite variable [4]. Pre-contact interactions between medusae and prey may influence capture probability but are poorly understood. This contrasts sharply with the fluid mechanics of propulsion by swimming medusae, which are well described [5–8]. The response patterns of prey to the fluid flows generated by medusae require quantification in order to understand how these responses affect the likelihood of prey capture.

Most scyphomedusan taxa generate similar hydrodynamic structures around their bells and in their wakes [9,10] but can differ considerably in the morphology of their capture surfaces [3]. Semaeostomes possess thin, long trailing tentacles that ring the bell margin, and broad oral arms surrounding the mouth (Figure 1a,b). In contrast, rhizostomes characteristically possess thick, dense oral disks that surround the mouth of the medusa (Figure 2c,d). Despite their diverse morphologies, all scyphomedusae swim similarly via rowing propulsion [10]. That is, they contract and relax their bells rhythmically. This cycle of contraction and relaxation generates large starting and stopping vortices, respectively, that circulate fluid through the capture structures–i.e., tentacles and oral arms for semaeostomes and digitata on the oral disks of rhizostomes.

The similarity of flow patterns among scyphomedusae suggests that differences in prey selection between the medusan orders are likely due to variations in prey contact with morphologically different tentacle structures. Semaeostomes tend to consume relatively large but slowly swimming zooplankton [11], while rhizostomes tend to consume smaller zooplankton such as rotifers and small larvae [12–14]. One past prediction was that most taxa of rowing jellyfish would not be good at capturing copepods because copepods swim and escape much faster than the feeding current generated by scyphomedusae [15]. Yet, copepods are often an important part of the gut contents of many scyphomedusae [11,16]. The source of copepod vulnerability to entrainment by such relatively slow feeding currents remains unclear and underscores our low knowledge about pre-capture events in the feeding process.

In an effort to understand why copepods are vulnerable to predation by scyphomedusae we quantified the pre-capture events that determine prey selection. Specifically, we quantified copepod prey interactions with feeding currents generated by two semaeostomes species (*Aurelia aurita* and *Chrysaora quinquecirrha*) and two rhizostomes species (*Catostylus tagi*, *Phyllorhiza punctata*) in order to understand how these interactions influence copepod capture success. All four species of medusa were fed the copepod *Acartia tonsa*. *Acartia tonsa* is known to be sensitive to hydrodynamic disturbances and capable of rapid escape swimming [17]. The capture efficiency of copepods by medusae was quantified in conjunction with transport paths of copepods in the feeding current and their reactive behavior.

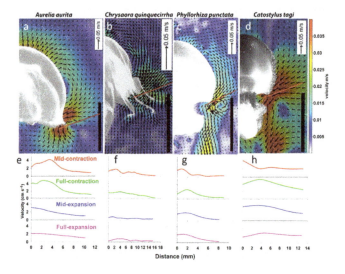

Figure 1. Velocity fields (**a**–**d**) and profiles (**e**–**h**) of representative individuals from two semeostomes species, *Aurelia aurita* and *Chrysaora quinquecirrha* and two rhizostome species *Phyllorhiza punctata* and *Catostylus tagi*. The velocity fields are of the flow adjacent to the bells at full-contraction and the velocity profiles show the velocities along a transect (red lines in **a**–**d**) that originates at the bell margin and extends away from the bell through the region that copepod prey are entrained in the feeding current. The profiles show how the velocity changes during different times throughout the swimming cycle.

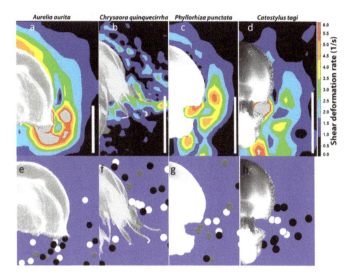

Figure 2. Maximum shear deformation fields around representative individuals ($n = 1$) from two semeostomes species, *Aurelia aurita* (**a**) and *Chrysaora quinquecirrha* (**b**) and two rhizostome species *Phyllorhiza punctata* (**c**) and *Catostylus tagi* (**d**). Corresponding shear fields illustrate shear at full-contraction. The black region is the region of shear below 1 s^{-1} a common detection threshold of copepod (*Acartia tonsa*) prey. Above that, many copepods should be capable of detecting the feeding current. (**e–h**) Locations relative to the medusan bell where copepods first jumped in reaction to feeding current. Each dot represents a different copepod and the different colors (white, grey, black) indicate the different replicated medusae used for quantifying predator-prey interactions ($n = 3$ different individuals per species).

2. Materials and Methods

All of the medusae (See Supplementary Table S1 for sizes) were supplied by the New England Aquarium in Boston, Massachusetts, USA. The medusae were housed in 37 L planktokreisel aquariums (Envision Acrylics Inc.) at their ambient temperatures. The medusae were all maintained at 12 h light and dark cycles. All medusae were starved one to two days before the trials began [3].

Hydrodynamic measurements (See Supplementary Table S1 for Reynolds Numbers based on velocities) were made using 2D digital particle image velocimetry (DPIV) following the methods of Colin et al. (2010). Glass filming vessels containing individual medusae were illuminated using a laser sheet (680-nm wavelength) directed through the central axis of the bell. The correct alignments of the laser light sheet was identified when the manubrium was fully illuminated, and only sequences where the medusa swam perpendicular to the screen were selected for analysis. This ensured that there was no motion in the unmeasured third dimension. A high-speed digital video camera (Fastcam 1024 PCI; Photron) was placed perpendicular to the light sheet to video record the individuals swimming at 500 frames s^{-1}. Flow velocity was determined from sequential images analyzed by a cross-correlation algorithm (LaVision Software). This analysis generated velocity vector fields around the medusa.

Each medusa's fluid interactions were quantified at four instances during the pulsation cycle. The first was at full bell relaxation and the fourth was at full bell contraction. The second and third instances corresponded to two instances equally spaced in time between the relaxed and contracted states.

The velocity vector fields were used to measure the four components of 2D shear deformation rates (rate of strain) E_{xx}, E_{xy}, E_{yx}, E_{yy} in the flow field, calculated with the equation set below

$$E_{xx} = \frac{du_x}{dx}$$
$$E_{xy} = \frac{du_y}{dx}$$
$$E_{yx} = \frac{du_x}{dy}$$
$$E_{yy} = \frac{du_y}{dy}$$
(1)

where u was the measured planar velocity field (relative to the observer). The maximum of these components was used to represent the maximum shear deformation rate, since a copepod prey item will elicit an escape response when the deformation rate is greater than its threshold regardless of its direction [17,18].

In order to visualize the predator-prey interactions, each medusa (n = 3 individuals per species) was individually filmed in an uncovered triangular glass tank (21 × 15 × 15 cm) with a mirror placed along the hypotenuse (Supplementary Figure S1). This arrangement enabled us to track copepod prey 3-dimensionally around the medusae [3,19]. The vessel was illuminated using LED lights and a camera (SONY HVR-77U Digital HD Video Camera Recorder, 30 frames per second) equipped with a standard lens (50 mm Nikon) video recorded the pulsing of the medusa 3-dimensionally.

The medusae were rigidly tethered in the the center of the vessel (about 5 cm from the surface and 15 cm from the bottom) by gluing the apex of the bell to a glass pipette using Superglue® [3,20]. Tethering does not alter the wake dynamics but it does shift the location from which the wake fluid originates. For untethered medusae, the wake fluid primarily originates from in front of the medusae, while for tethered medusae it originates from the side (unpublished data). A ruler was placed in the vessel at the beginning of each recording for the reference scale. For each individual, video recording commenced after 5 to 20 min when 'normal' swimming behavior was observed consistently for several minutes (i.e., 5–10 min). Live prey of wild *Acartia tonsa* (male/female ranging from 0.9 to 1.1 mm) were introduced into the filming vessel in high enough concentration to initiate medusan feeding (~0.5 mL^{-1}). Each feeding medusa was recorded individually for 30 to 45 min.

To quantify the interaction between copepod prey and medusae, encounters of copepods with each medusa was defined as copepods visibly being entrained in the flow of the medusa. Therefore, copepods were chosen for tracking by observing near the medusa (X, Y) until a copepod was seen to move swiftly in the feeding current (defined as entrainment). Entrainment was easy to identify because when copepods were not entrained in the feeding current they remained stationary except when they jumped. Upon recognition of an encounter, the video was then played forward to observe the outcome of the interaction, then played in reverse until the copepod first became entrained in the flow. This ensured a clear recording of the full track of the copepod during entrainment. The copepod position was tracked over time relative to the bell with the distance from the bell being determined using the X, Y Z coordinates of the copepod.

The copepod trajectories were digitized by hand and analyzed using ImageJ (NIH) software. Tracking occurred every 5 frames, starting at the initial time of the encounter (i.e., when it started to be entrained). The copepod's position was recorded every 5 frames. Copepods were tracked until the copepod either escaped (jumped or transported out of the flow) or was consumed. Since medusae are axisymmetric, all tracks were transposed to one side of the medusae to enable visualization of how the tracks overlapped relative to the bell margin.

Statistical analyses were performed using Sigma Plot® statistical software. Analysis of variance tests were performed to compare capture efficiencies among medusan species (n = 3 different individuals per species). Comparisons were made among the 4 medusan species. Holm-Sidak tests were used to make post-hoc comparisons to compare efficiencies between the semaeostome and rhizostome species (significance level, α = 0.05). All of the data conformed to the assumptions of homoscedasticity (Browne-Forsythe test) and normality (Shapiro-Wilk test).

3. Results

To initiate encounters with prey, scyphomedusae pulse by cyclically contracting and relaxing their bells. These bell motions entrained fluid along the bell margin where flow velocities around the medusae peaked (Figure 1). Maximum flow velocities around all the species studied were less than 5 cm s^{-1} (Figure 1) and decreased with distance from the bell margin (measured along a transect in the direction of the primary path of entrainment; Figure 1e–h). Entrainment velocities peaked during bell contraction but fluid was entrained toward the bell throughout the swimming cycle (Figure 1e–h).

Velocity fields resulted in fluid shear deformation rates (the primary fluid property to which copepods respond) considerably greater than 1 s^{-1} surrounding the medusan bells (seen as any colored regions in Figure 2a–d and above red dashed lines in Figure 3). The copepod *Acartia tonsa* is capable of detecting shear rates >0.5 s^{-1} [17] so they should be able to sense the feeding current somewhere in the black regions in Figure 2a–d. Like velocity, shear rates were greater during bell contraction but were above the detection threshold of *A. tonsa* within 5 mm from the bell margin (which is equivalent to 5 copepod body lengths) throughout the swimming cycle. The patterns seen for the velocity and shear fields were similar among the four medusan species examined.

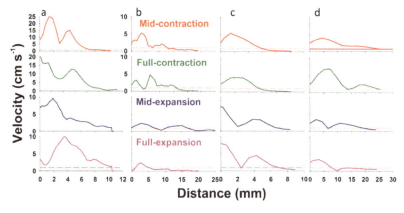

Figure 3. Maximum shear deformation profiles along the transect (red line) shown in Figure 1a–d throughout the swimming cycle for two semeostomes species, *Aurelia aurita* (**a**) and *Chrysaora quinquecirrha* (**b**) and two rhizostome species *Phyllorhiza punctata* (**c**) and *Catostylus tagi* (**d**). The dashed red line shows the 1 s^{-1} shear level which is a common detection threshold of copepod prey.

Tracks of copepods entrained in the flow around the swimming medusae showed that copepod encounters occurred when copepods located in front or to the side of the medusae were transported toward the bell margin (Figure 4). Upon entrainment, copepods were transported toward the capture surfaces of medusae—i.e., tentacles and oral arms for semaeostomes and oral disk for rhizostomes. The raw tracks illustrate that most of the copepods encountered by the medusae (encounter defined as a copepod entrained in the flow) were not captured (grey tracks) and were ultimately transported downstream away from the medusae. The location of the tracks of the captured copepods (red) overlapped with the tracks for the copepods not captured, demonstrating that captured prey do not originate from a particular part of the flow. Of the copepods entrained in the flow, the proportion that reacted to the flow varied with 78% reacting to *Aurelia aurita* and only 38% reacting to *Catostylus tagi* (Supplementary Table S2). A comparison of the locations where the copepods first jumped to the fluid shear fields (Figure 2) suggest that most the copepods did not react until shear rates well above the documented threshold of *Acartia tonsa* (approximately 0.5 s^{-1}). Copepods reacted with either a small adjustment jump of a distance of 1.1 ± 0.1 mm (about 1 body length) or a stronger escape jump of a distance of 5.1 ± 1.4 mm (about 5 body lengths; Supplementary Table S2).

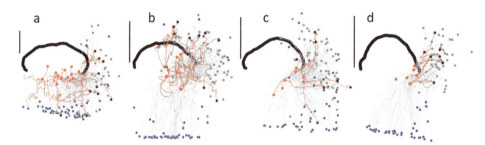

Figure 4. Tracks of entrained copepods during the encounter process with two semeostomes species, *Aurelia aurita* (**a**) and *Chrysaora quinquecirrha* (**b**) and two rhizostome species *Phyllorhiza punctata* (**c**) and *Catostylus tagi* (**d**). Grey tracks are the tracks of copepods that escaped. Dark grey and blue dots indicates their starting and ending position during tracking. Red tracks are tracks of captured copepods. Black dots indicate starting location and red dots indicate capture location.

Overall, only 18.0% (±6.8) of the copepods entrained in the flow of the scyphomedusae studies were ultimately captured. The capture efficiencies ranged from 23.5% for *A. aurita* to 13.9% for *C. tagi*, however, mean capture efficiencies did not differ significantly among the four species (Figure 5a; ANOVA, $n = 3$ medusae, $F_{3,7} = 1.39$, $P = 0.32$). Surprisingly, the behavior of the copepods did not significantly impact capture efficiencies (Figure 5b; ANOVA, $n = 3$ medusae, $F_{2,6} < 0.48$, $P > 0.6$ for all comparisons). Figure 6 illustrates the cumulative outcome of encounters with copepod prey for each of the medusan species examined (and cumulatively for all species). Following the thicker blue flow lines for the semeostomes *A. aurita* and *C. quinquecirrha* shows that most of the copepods reacted to entrainment but that most only reacted with a few small adjustment jumps and were not captured. For the rhizostomes *P. punctata* and *C. tagi*, most of the copepods did not react to encounters. Overall for all species, a comparison of the percentages in the green (not captured) and red (captured) boxes illustrates that the fate of encountered copepods was not dramatically different for the different behaviors (no reaction, reposition jump or escape jump).

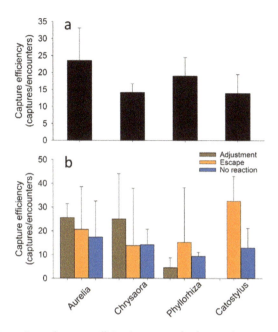

Figure 5. (a) Comparison of capture efficiencies among the four medusan species *Aurelia aurita*, *Chrysaora quinquecirrha*, *Phyllorhiza punctata* and *Catostylus tagi*. fed *Acartia tonsa* copepods. (b) Comparison of capture efficiencies of copepods with different behaviors after entrainment in the medusan feeding currents.

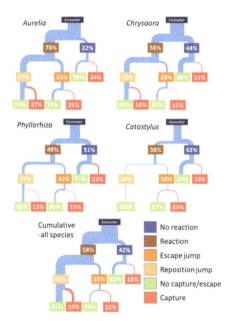

Figure 6. Flow diagrams illustrating the cumulative outcomes of all of the copepod encounters with each of the medusan species *Aurelia aurita* (n = 78 copepod encounters), *Chrysaora quinquecirrha* (n = 84), *Phyllorhiza punctata* (n = 61) and *Catostylus tagi* (n = 48) and cumulative among all species (n = 271).

4. Discussion

The feeding current generated by rowing scyphomedusae transports large volumes of fluid through the medusa's capture surfaces [9]. The volume of transported fluid is large enough that medusa populations can process all the water in a coastal embayment within hours [21–24]. Bulk processing of fluid is necessary because these medusae depend upon chance contact of entrained prey with trailing medusan capture surfaces. Consequently, our study (as well as others [3,4,25]) indicates that prey capture efficiency is low (<25%; Figure 4).

The hydrodynamic characteristics of the feeding current generated by scyphomedusae differs considerably from that of the other important predators of copepods—i.e., fish and ctenophores. The feeding current of lobate ctenophores is highly effective at capturing copepods [26,27]. That feeding current is a slow, laminar flow characterized by fluid shear deformation rates below levels that copepods are capable of detecting [26]. Furthermore, lobate ctenophores scan incoming flows and redirect the feeding current (and prey) rather than relying upon random contacts of prey with capture surfaces [19]. As a result, the lobate ctenophore *Mnemiopsis leidyi* is capable of capturing >85% of the copepods entrained in its feeding current [26,27]. Fish also rely on a stealthy approach and an added rapid strike with high speed suction to capture copepods [28,29]. This strategy enables fish to capture nearly 90% of the copepods they attack [30,31]. Despite this high success rate, copepod escape behavior is an important influence on the capture success of fish [31–33].

In contrast, scyphomedusan feeding currents are neither stealthy nor very fast. We show that the maximum fluid velocity in the feeding current of the scyphomedusae examined is about 4 cm s^{-1} (Figure 1). Feeding current velocities of scyphomedusae are therefore substantially slower than the escape velocities of copepods. If copepods are capable of detecting the feeding current, then scyphomedusae would not be expected to readily capture copepods [15,34]. Our data show that the fluid shear deformation rate of the feeding current is well above the detection limit of *A. tonsa* even at distance >5 mm from the medusae (which is >5 copepod body-lengths; Figures 2 and 3). Therefore, copepods should be capable of detecting the medusan flows before the copepods are transported in the feeding current to a medusan capture surface. However, as we have shown, scyphomedusae readily capture a fraction of *A. tonsa* and other copepods [11,16].

The behavior of copepods in scyphozoan feeding currents indicate several factors that could affect capture of copepods by medusae. First, among all the medusan species studied, 34% of the copepods did not react while entrained in the feeding current. In fact, for the rhizostome medusae >50% of the copepods did not react throughout the encounter process. This percentage of non-responsive copepods is relatively high where other studies have shown that 17% of *Temora turbinata* and 10% *A. hudsonica* did not respond [35,36]. However, juvenile *A. hudsonica* copepodites were much less responsive with >50% not responding to the medusan feeding current [35]. Second, of those that did react, most reactions were small adjustment jumps (84% among all species) rather than strong escape jumps. These jumps did not transport the copepods outside the flow and most of those copepods were transported through trailing capture surfaces such as tentacles or oral arms. Most surprisingly, the behavior of the copepods–no response, mild jump, or even a strong jump-did not have a significant overall effect on the overall capture efficiencies. The probability of a copepod being captured remained around 18% regardless of copepod behavior.

The observation that copepod behavior did not strongly impact capture efficiency is counter-intuitive. However, studies that quantified fluid transport in the feeding current of medusae using finite-time Lyapunov exponent (FTLE) and Lagrangian coherent structures (LCS) demonstrated scyphozoan feeding currents to be a complex of fluid packets, or lobes (Figure 7). Only a limited portion of the flow is comprised of fluid packets that directly impinge on capture surfaces. Further, within the packets that contact capture surfaces, copepod behavior only modestly affects the encounter outcome [37]. Assuming that only prey in these packets can be caught, these studies have shown that prey in these components of the feeding current may be transported to capture surfaces, despite escape jumps (Figure 7). These studies also show that these packets of 'captured fluid' originate from locations adjacent to packets of non-captured fluid (Figure 7). Therefore, unresponsive behavior by copepods within these flow regions may only modestly influence the copepod's chance of survival [37]. This is consistent with prey movement patterns (Figure 4) demonstrating that the tracks of captured copepods overlap with the tracks of non-captured copepods. When these LCS studies incorporated copepod escape swimming in response to shear (>1 s^{-1}) they found that copepod escape jumping reduced captures by <10% (Figure 7). The picture that emerges from these considerations is one in which the chance location of a copepod within distinct components of the medusan flow, rather than the flow's average velocity, may be the primary determinant of capture after contact. Since these regions of high capture probability comprise only a limited portion of the overall flow, the potential for prey to contact capture surfaces is limited and escape behavior may be a modest component of prey escape probability.

Figure 7. Lagrangian coherent structures (LCS) and capture regions for prey with different perception threshold values. Threshold = 0 (constant escape response, white line), threshold = 1 s^{-1} (red line), no escape response (green line). From Peng and Dabiri 2009 (with permission).

5. Conclusions

While semeostome and rhizostome medusae lack the sensory capabilities and stealth of ctenophores or fish, the feeding current of medusae is large and hydrodynamically complex [4,6,9,37]. The scale and complexity of the feeding current may explain why copepod escape behavior has only a modest effect on capture efficiencies (Figure 5; [3,36,38]), enabling medusae to capture a proportion of the copepods entrained in the feeding current. While scyphomedusae are capable of capturing copepods, the inefficiency of their feeding mechanics result in low overall capture efficiencies and limits their trophic impact. Their feeding strategy, characterized by low capture efficiencies offset by bulk fluid processing, ultimately results in scyphomedusae being capable of significantly impacting prey standing stocks only when medusan populations achieve peak abundances [10,39] and references therein.

Supplementary Materials: The following are available online at http://www.mdpi.com/2311-5521/5/2/60/s1, Figure S1: Schematic of 3-dimensional video set-up. The tethered medusae were placed into the right-triangle shaped filming vessel (21 × 15 × 15 cm) with the hypotenuse side constructed using a mirror, Table S1: Sizes (n = 3 individuals per species) and Reynolds Numbers (based on bell diameter and maximum flow velocity) of the four scyphomedusan species used in the study, Table S2: Reaction of copepods to different scyphomedusae. The number and percent of copepods that did not react or made a repositioning jump or an escape jump. Average distance (n = 3 individuals per species) that the copepods moved for each type of jump.

Author Contributions: Author contributions: methodology, S.P.C., J.H.C.; validation, S.P.C., J.H.C.; formal analysis, Z.W., S.P.C.; writing—original draft preparation, Z.W., S.P.C.; writing—review and editing, S.P.C., J.H.C.; visualization, S.P.C.; supervision, S.P.C.; funding acquisition, S.P.C., J.H.C. All authors have read and agreed to the published version of the manuscript.

Funding: This research was funded by National Science Foundation Biological Oceanography grant awarded to S.P.C. and J.H.C. (OCE 1536688 and OCE 1536672) and supported by Roger Williams University Foundation to Promote Teaching and Scholarship.

Conflicts of Interest: The authors declare no conflict of interest.

References

1. Behrends, G.; Schneider, G. Impact of Aurelia aurita medusae (Cnidaria, Scyphozoa) on the standing stock and community composition of mesozooplankton in the Kiel Bight (western Baltic Sea). *Mar. Ecol. Prog. Ser.* **1995**, *127*, 39–45. [CrossRef]
2. Graham, W.M.; Gelcich, S.; Robinson, K.L.; Duarte, C.M.; Brotz, L.; Purcell, J.E.; Madin, L.P.; Mianzan, H.; Sutherland, K.R.; Uye, S.I.; et al. Linking human well-being and jellyfish: Ecosystem services, impacts, and societal responses. *Front. Ecol. Environ.* **2014**, *12*, 515–523. [CrossRef]
3. Bezio, N.; Costello, J.H.; Perry, E.; Colin, S.P. Effects of capture surface morphology on feeding success of scyphomedusae: A comparative study. *Mar. Ecol. Prog. Ser.* **2018**, *596*, 83–93. [CrossRef]
4. Katija, K.; Beaulieu, W.T.W.; Regula, C.; Colin, S.S.P.; Costello, J.H.J.; Dabiri, J.O. Quantification of flows generated by the hydromedusa Aequorea victoria: A Lagrangian coherent structure analysis. *Mar. Ecol. Prog. Ser.* **2011**, *435*, 111–123. [CrossRef]
5. Dabiri, J.O.J.O.; Gharib, M.; Colin, S.P.; Costello, J.H. Vortex motion in the ocean: In situ visualization of jellyfish swimming and feeding flows. *Phys. Fluids* **2005**, *17*, 091108. [CrossRef]
6. Gemmell, B.J.; Troolin, D.R.; Costello, J.H.; Colin, S.P.; Satterlie, R.A. Control of vortex rings for manoeuvrability. *J. R. Soc. Interface* **2015**, *12*, 20150389. [CrossRef]
7. Colin, S.P.S.P.; Costello, J.H.; Dabiri, J.O.J.O.; Villanueva, A.; Blottman, J.B.J.B.; Gemmell, B.J.B.J.; Priya, S. Biomimetic and live medusae reveal the mechanistic advantages of a flexible bell margin. *PLoS ONE* **2012**, *7*, e48909. [CrossRef]
8. Gemmell, B.J.B.J.; Costello, J.H.; Colin, S.P.; Stewart, C.J.; Dabiri, J.O.; Tafti, D.; Priya, S. Passive energy recapture in jellyfish contributes to propulsive advantage over other metazoans. *Proc. Natl. Acad. Sci. USA* **2013**, *110*, 17904–17909. [CrossRef]
9. Dabiri, J.O.; Colin, S.P.; Costello, J.H.; Gharib, M. Flow patterns generated by oblate medusan jellyfish: Field measurements and laboratory analyses. *J. Exp. Biol.* **2005**, *208*, 1257–1265. [CrossRef]
10. Costello, J.H.; Colin, S.P.S.P.; Dabiri, J.O.J.O. Medusan morphospace: Phylogenetic constraints, biomechanical solutions, and ecological consequences. *Invertebr. Biol.* **2008**, *127*, 265–290. [CrossRef]
11. Graham, W.M.; Kroutil, R.M. Size-based Prey Selectivity and Dietary Shifts in the Jellyfish, Aurelia aurita. *J. Plankton Res.* **2001**, *23*, 67–74. [CrossRef]
12. Larson, J. Diet, Prey Selection and Daily Ration of Stomolophus a Filter-feeding Scyphomedusa from the NE Gulf of Mexico. *Estuar. Coast. Shelf Sci.* **1991**, *32*, 511–525. [CrossRef]
13. Peach, M.B.; Pitt, K.A. Morphology of the nematocysts of the medusae of two scyphozoans, Catostylus mosaicus and *Phyllorhiza punctata* (Rhizostomeae): Implications for capture of prey. *Invertebr. Biol.* **2007**, *124*, 98–108. [CrossRef]
14. Álvarez-Tello, F.J.; López-Martínez, J.; Lluch-Cota, D.B. Trophic spectrum and feeding pattern of cannonball jellyfish Stomolophus meleagris (Agassiz, 1862) from central Gulf of California. *J. Mar. Biol. Assoc. UK* **2016**, *96*, 1217–1227. [CrossRef]
15. Costello, J.H.; Colin, S.P. Morphology, fluid motion and predation by the scyphomedusa Aurelia aurita. *Mar. Biol.* **1994**, *121*, 327–334. [CrossRef]
16. Suchman, C.L.; Sullivan, B.K. Vulnerability of the copepod Acartia tonsa to predation by the scyphomedusa Chrysaora quinquecirrha: Effect of prey size and behavior. *Mar. Boil.* **1998**, *132*, 237–245. [CrossRef]
17. Kiorboe, T.; Saiz, E.; Visser, A.W. Hydrodynamic signal perception in the copepod Acartia tonsa. *Mar. Ecol. Prog. Ser.* **1999**, *179*, 97–111. [CrossRef]
18. Kiorboe, T.; Visser, A.W. Predator and prey perception in copepods due to hydromechanical signals. *Mar. Ecol. Prog. Ser.* **1999**, *179*, 81–95. [CrossRef]

19. Colin, S.P.; MacPherson, R.; Gemmell, B.; Costello, J.H.; Sutherland, K.; Jaspers, C. Elevating the predatory effect: Sensory-scanning foraging strategy by the lobate ctenophore Mnemiopsis leidyi. *Limnol. Oceanogr.* **2015**, *60*, 100–109. [CrossRef]
20. Corrales-Ugalde, M.; Colin, S.P.; Sutherland, K.R. Nematocyst distribution corresponds to prey capture location in hydromedusae with different predation modes. *Mar. Ecol. Prog. Ser.* **2017**, *568*, 101–110. [CrossRef]
21. Olesen, N.J. Clearance potential of jellyfish Aurelia aurita, and predation impact on zooplankton in a shallow cove. *Mar. Ecol. Prog. Ser.* **1995**, *124*, 63–72. [CrossRef]
22. Olesen, N.; Frandsen, K.; Riisgård, H.U. Population dynamics, growth and energetics of jellyfish Aurelia aurita in a shallow fjord. *Mar. Ecol. Prog. Ser.* **1994**, *105*, 9–18. [CrossRef]
23. Bay, C.; Purcell, J.E. Effects of predation by the scyphomedusan Chrysaora quinquecirrha on zooplankton. *Mar. Ecol. Prog. Ser.* **1992**, *87*, 65–76.
24. Graham, E.S.; Tuzzolino, D.M.; Burrell, R.B.; Breitburg, D.L. Interannual Variation in Gelatinous Zooplankton and Their Prey in the Rhode River, Maryland. *Smithson. Contrib. Mar. Sci.* **2009**, *38*, 369–377.
25. Riisgård, H.U.; Madsen, C.V. Clearance rates of ephyrae and small medusae of the common jellyfish Aurelia aurita offered different types of prey. *J. Sea Res.* **2011**, *65*, 51–57. [CrossRef]
26. Colin, S.P.; Costello, J.H.; Hansson, L.J.; Titelman, J.; Dabiri, J.O. Stealth predation and the predatory success of the invasive ctenophore Mnemiopsis leidyi. *Proc. Natl. Acad. Sci. USA* **2010**, *107*, 1–5. [CrossRef]
27. Waggett, R.J.; Costello, J.H. Capture mechanisms used by the lobate ctenophore, Mnemiopsis leidyi, preying on the copepod Acartia tonsa. *J. Plankton Res.* **1999**, *21*, 2037–2052. [CrossRef]
28. Holzman, R.; Wainwright, P.C. How to surprise a copepod: Strike kinematics reduce hydrodynamic disturbance and increase stealth of suction-feeding fish. *Computer* **2009**, *54*, 2201–2212. [CrossRef]
29. Gemmell, B.J.; Adhikari, D.; Longmire, E.K. Volumetric quantification of fluid flow reveals fish's use of hydrodynamic stealth to capture evasive prey Volumetric quantification of fluid flow reveals fish's use of hydrodynamic stealth to capture evasive prey. *J. R. Soc. Interface* **2014**, *11*, 20130880. [CrossRef]
30. Coughlin, D.J.; Strickler, J.R. Zooplankton capture by a coral reef fish: An adaptive response to evasive prey. *Environ. Biol. Fishes* **1990**, *29*, 35–42. [CrossRef]
31. Gemmell, B.J.; Buskey, E.J. The transition from nauplii to copepodites: Susceptibility of developing copepods to fish predators. *J. Plankton Res.* **2011**, *33*, 1773–1777. [CrossRef]
32. Waggett, R.J.; Buskey, E.J. Copepod escape behavior in non-turbulent and turbulent hydrodynamic regimes. *Mar. Ecol. Prog. Ser.* **2007**, *334*, 193–198. [CrossRef]
33. Waggett, R.J.; Buskey, E.J. Calanoid copepod escape behavior in response to a visual predator. *Mar. Biol.* **2007**, *150*, 599–607. [CrossRef]
34. Costello, J.H.; Colin, S.P. Flow and feeding by swimming scyphomedusae. *Mar. Biol.* **1995**, *124*, 399–406. [CrossRef]
35. Suchman, C.L. Escape behavior of Acartia hudsonica copepods during interactions with scyphomedusae. *J. Plankton Res.* **2000**, *22*, 2307–2323. [CrossRef]
36. Nagata, R.M.; Morandini, A.C.A.; Colin, S.P.; Migotto, A.E.; Costello, J.H.J. Transitions in morphologies, fluid regimes, and feeding mechanisms during development of the medusa *Lychnorhiza lucerna*. *Mar. Ecol. Prog. Ser.* **2016**, *557*, 145–159. [CrossRef]
37. Peng, J.; Dabiri, J.O. Transport of inertial particles by Lagrangian coherent structures: Application to predator–prey interaction in jellyfish feeding. *J. Fluid Mech.* **2009**, *623*, 75–84. [CrossRef]
38. Suchman, C.L.; Sullivan, B.K. Effect of prey size on vulnerability of copepods to predation by the scyphome *Aurelia aurita* and *Cyanea* sp. *J. Plankton Res.* **2000**, *22*, 2289–2306. [CrossRef]
39. Purcell, J.E.; Arai, M.N. Interactions of pelagic cnidarians and ctenophores with fish: A review. *Hydrobiologia* **2001**, *451*, 27–44. [CrossRef]

© 2020 by the authors. Licensee MDPI, Basel, Switzerland. This article is an open access article distributed under the terms and conditions of the Creative Commons Attribution (CC BY) license (http://creativecommons.org/licenses/by/4.0/).

Article

Feeding of Plankton in a Turbulent Environment: A Comparison of Analytical and Observational Results Covering Also Strong Turbulence

Hans L. Pécseli [1],*, Jan K. Trulsen [2], Jan Erik Stiansen [3] and Svein Sundby [3]

[1] Department of Physics, University of Oslo, N-0316 Oslo, Norway
[2] Institute of Theoretical Astrophysics, University of Oslo, N-0315 Oslo, Norway; j.k.trulsen@astro.uio.no
[3] Institute of Marine Research, N-5817 Bergen, Norway; jan.erik.stiansen@hi.no (J.E.S.); svein.sundby@hi.no (S.S.)
* Correspondence: hans.pecseli@fys.uio.no

Received: 23 January 2020; Accepted: 10 March 2020; Published: 19 March 2020

Abstract: The present studies address feeding of plankton in turbulent environments, discussed by a comparison of analytical results and field data. Various models for predator-prey encounters and capture probabilities are reviewed. Generalized forms for encounter rates and capture probabilities in turbulent environments are proposed. The analysis emphasizes ambush predators, exemplified by cod larvae *Gadus morhua* L. in the start-feeding phase (stage 7 larvae) collected in shallow waters near Lofoten, Norway. During this campaign, data were obtained at four sites with strongly turbulent conditions induced by tidal currents and long-wave swells, and one site where the turbulence had a lower level in comparison. The guts of the selected cod larvae were examined in order to determine the number of nauplii ingested. Analytically obtained probability densities for the gut content were compared with observations and the results used for estimating the rate of capture of the nauplii. This capture rate was then compared with analytical results using also data for the surroundings, such as measured prey densities and turbulence conditions, as quantified by the specific energy dissipation rate. Different from earlier studies, the presented data include conditions where the turbulence exceeds the level for optimal larval encounter-capture rates.

Keywords: plankton; turbulence; data analysis

---

1. Introduction

The survival rates of fish in their early life-stages are influenced by a number of biological and physical processes [1,2]. A number of models have been proposed for several individual partial processes. Models for predator-prey encounter rates, in particular, form the basis of many biological applications, including the feeding rates of larval fish and the implications of environmental effects on their growth and survival [3,4]. Pioneering studies [5,6] argued that turbulent motions in the local environment could be important by enhancing encounter rates between predators and prey. Prior to these studies, turbulence was assumed to reduce predator-prey encounter rates because of the effect of prey dilution and breaking down the peak concentrations. The question was how known physical laws for turbulence modeling, such as the Kolmogorov-Obukhov law for the second order structure function, are reflected in the predator-prey encounter rate. Many elements of various models have been tested under controlled laboratory conditions using small polystyrene particles as representing plankton [7,8]. The turbulence conditions could be varied systematically by controlling the external stirring mechanism that created the turbulence. Models were tested also by numerical simulations where plankton was represented by point particles passively following the turbulent motions in the flow [9,10]. Studies allowing for self-induced motions have also been presented [11,12]. It was, however, realized that too strong turbulence could have an adverse effect for the predator by reducing the capture probability of prey [13–15]. Low turbulence levels give an advantage by increasing the encounter rate, while too strong turbulence can be disadvantageous by reducing the capture rate to such an extent that it partially dominates the effect of turbulent encounters. The result is the existence of an optimum turbulence level that is a characteristic for the given species. A dome-shaped relationship for the predator capture rate as a function of turbulence intensity ϵ has been suggested based on predator response time and and reactive distance [13]. The present study outlines the properties of various models for the encounter rate in turbulent environments and includes also the consequences of the reduce capture rate for large turbulence levels. For ambush predators we assume that the encounter and capture rates are basically controlled by turbulent motions in the environment. For the capture they can rely on both visual identification and organs sensing disturbances in the water [16,17].

In the present study we compare analytical results with results obtained by analyzing field data of the gut content of fish larvae collected in the shallow waters near Lofoten, Norway. Additional material is presented elsewhere [18]. Predators (in the present case cod larvae, *Gadus morhua* L.) were sampled by two different methods: 1) an ichthyoplankton net moved slowly from the seabed to the surface, thereby vertically integrating the larval samples, and 2) a high-capacity submersible pump (HUFSA) at selected positions 5, 10, 15, 20, 25 and 30 m below the surface. The guts of the selected larvae were opened under a microscope and a database established for the number of nauplii in the guts, together with the length of the corresponding cod larvae. The entire database consists here of 3247 entries, containing the simultaneously obtained lengths and gut contents of cod larvae supported by the experimentally determined local specific energy dissipation rate ϵ. The data obtained by the pump are depth resolved, with corresponding local prey concentrations. The subset collected by the net is an average over all depths, and this part of the database contains 866 entries. Cruising speeds of cod larvae as observed during laboratory conditions are in the velocity range 0.1–0.3 mm s^{-1} [19]. This can be approximately one order of magnitude higher cruising speed than that of their naupliar prey [20]. We do not expect any significant differences between the two datasets as far the vertical distribution is concerned. For a small subset consisting of 299 samples evenly distributed over depth, also the lengths of the nauplii found in the guts were measured. The ambient concentrations of the nauplii (prey) were sampled by a zooplankton pump at discrete depth positions. The details of the biological sampling in the field and the processing of the samples in the laboratory is similar to two other studies [21,22]. These studies deduced the cod larval ambient turbulence energy dissipation rate of the mixed layer from averaged wind-induced turbulence based on the empirical results [23]. The present study, on the other hand, calculated the turbulent energy dissipation rate ϵ by measuring the turbulence using acoustic current meters from a point observation at the top of a 6.5 m high tower, deployed on the seabed at the shallow observation sites.

The database obtained on the basis of the field data were used for a test of analytical predictions. Analytical models for probability densities for the gut content were thus compared with observations to give an estimate of the rate of capture of the nauplii. This capture rate was then compared with analytical results.

2. Results

The present study contains two parts, analytical results and study of data from a field experiment, where the data could be analyzed to give a form amenable for comparison with theory. The present section is divided into two parts addressing these two approaches separately.

2.1. Analytical Results

The present analytical part will mostly be a review of some of the existing models for predator-prey encounter rates and capture probabilities in turbulent environments. The discussion is separated into two parts.

2.1.1. Encounter Rates in Turbulent Flows

To illustrate the complexity of the problem of particle moving in a turbulent flow we show Figures 1 and 2. The two figures allow a three-dimensional, stereoscopic view. It requires a little exercise. The observer should focus the eyes approximately 20 cm behind the plane of the paper or computer screen. The distance to the eyes is not so critical provided it is sufficiently large, but it is essential that the figure is kept plane and horizontally aligned with the observers eyes. Figure 1 shows a group of point particles moving in a turbulent flow; one of the particles indicated by red color represents the predator, and the others a group of prey that happened to be close to it at some initial time. The presentation is given in the fixed laboratory, or Eulerian, frame. At first sight, the motion does not seem to be particularly complicated; the particles move like a group with some small relative motion. The problem is that the figure is shown in the wrong frame of reference. The correct frame for the present problem is the Lagrangian frame, one that moves with the predator. This is shown in Figure 2 and here the relative motions are much more complicated. A complete detailed analysis of the motion of the particles is not possible, and only a statistical analysis is feasible. The desired results should account for the average flux of particles presenting prey through a given surface (not necessarily spherical) that represents the range of the predator. We associate a characteristic length R_c with this range.

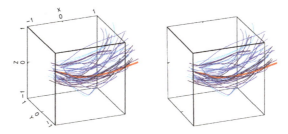

Figure 1. Allowing for a three dimensional, stereoscopic view, the figure shows the motion of a small cloud of selected particles moving in a turbulent flow. Units on axes are in computational units. The heavy red line shows the trajectories in an Eulerian reference frame for the reference predator. Time increases to the right. The figure is representative for the range R_c being in the inertial subrange. For comparison we have the scale size of the largest energy containing eddies to be ~ 3 in the present computational units. A possible self-induced motion of the predator is ignored here.

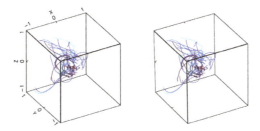

Figure 2. Trajectories for the point particles in Figure 1 here shown in the Lagrangian or co-moving frame for the reference predator represented by a central red point in this co-moving frame. The present figure as well as Figure 1 is based on data from numerical simulations [24,25].

Depending on the parameters of the turbulence and the characteristics of the predator, we can envisage two limiting cases. One is where R_c is in the inertial range of the turbulence, and one where it is in the viscous subrange. A dimensional analysis estimates the separation length as the Kolmogorov microscale $\eta = (\nu^3/\epsilon)^{1/4}$, where ν is the kinematic viscosity of the water [26]. A numerical coefficient can, however, not be determined by this argument, and it has to be found by other methods.

We consider the longitudinal velocity structure function $\langle (u_\|(\mathbf{x},t) - u_\|(\mathbf{x}+\mathbf{r},t))^2 \rangle / \sqrt{\nu\epsilon}$, where the subscript $\|$ indicates the velocity component in the direction along $\mathbf{r}$. The Kolmogorov velocity scale is $u_K = (\nu\epsilon)^{1/4}$. Two relevant subranges can be identified for the turbulent velocity fluctuations, (1) the inertial range with the Kolmogorov-Obukhov structure function $C_K \epsilon^{2/3} r^{2/3}$ with $C_K \approx 2.0 - 2.5$ being an empirically determined numerical coefficient [27], and (2) the viscous subrange with a structure function $C_\nu r^2 \epsilon / \nu$, where a numerical coefficient $C_\nu = 1/15$ is obtained analytically [28]. A modified Kolmogorov scale η_0 is defined as the scale separating these two universal subranges. It is readily determined as the length scale $\eta_0 \equiv \eta (15 C_K)^{3/4}$, where the values of the two structure functions are equal [10], see also Figure 3. We find $(15 C_K)^{3/4} \approx 13$ so it is not a trivial correction factor.

In Figure 3 we show by a full continuous line a phenomenological interpolation function for the structure function covering both inertial and viscous subranges [29]. The analytical form is

$$St(r/\eta) = \frac{C_K C_\nu (r/\eta)^{8/3}}{C_\nu (r/\eta)^2 + C_K (r/\eta)^{2/3}}. \tag{1}$$

With the given normalization used for the structure function in Figure 3, we find that $St(r/\eta)$ reproduces both the inertial and viscous subranges for large and small r, respectively.

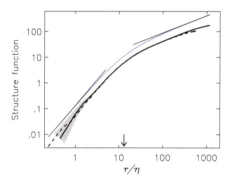

Figure 3. Numerically obtained normalized longitudinal structure function $\langle (u_\|(\mathbf{x},t) - u_\|(\mathbf{x}+\mathbf{r},t))^2 \rangle / u_K^2$ for the velocity component parallel to the separation vector $\mathbf{r}$, shown on a double logarithmic scale for varying normalized separation r/η. The reference velocity used for normalizations is the Kolmogorov velocity $u_K \equiv (\nu\epsilon)^{1/4}$. Results from two simulations are shown with full and dashed lines, respectively. Analytical results are given for the inertial and viscous subranges, by the slopes of the thin lines with $r^{2/3}$ and r^2, respectively. For clarity of presentation, the two slopes as well as the analytical approximation (1) have been offset vertically. The structure functions are uncertain for the smallest separations, as indicated by a shading. An vertical arrow indicates $r = \eta_0 \approx 13\eta$.

The encounter rate (here denoted J_e) for point particles in turbulent flows have been investigated analytically by a significant number of studies, noting that relevant results, apart from a numerical constant, can be obtained with a relatively straight forward dimensional reasoning [8]. These idealized point-particles are then assumed to represent predators and prey. The accuracy of this assumption has been discussed [10], suggesting some improvements for the modeling on the expense of more complicated analytical expressions. The full analysis needs not be reproduced here since it is available in the literature, where also reviews can be found [30]. We have an expression for the turbulent encounter rate for predators moving like point particles surrounded by a spherical encounter surface with a radius R_c in the inertial subrange. For this idealized case the expression

$$J_{e,I} = C_M D_0 \epsilon^{1/3} R_c^{7/3} \qquad (2)$$

was found [29], with C_M being a universal numerical coefficient. The expression (2) reproduces also previous results [6] for the $\epsilon^{1/3} R_c^{7/3}$ scaling in the inertial subrange $\eta_0 < R_c < L_E$, with L_E being the outer scale limiting the inertial subrange. We introduced D_0 as the reference density (or concentration) of prey. In the limit $R_c > \eta_0$, it is possible to obtain analytical expressions [31] for the full time variation of the flux when the problem is solved for an initial condition with constant density for $r > R_c$. For spherical encounter surfaces with radius R_c in the viscous subrange an expression in the form

$$J_{e,v} = C_v D_0 R_c^3 \sqrt{\frac{\epsilon}{\nu}} \qquad (3)$$

was found [29]. The expression (3) gives an $R_c^3 \sqrt{\epsilon/\nu}$ scaling of the turbulent flux for the viscous subrange $0 < R_c < \eta_0$, with a numerical constant $C_v \neq C_M$. From experiments or numerical simulations, the numerical coefficients were determined empirically [29] to be $C_v \approx 1$ while $C_M \approx 6-7$. The dimensionless combination $D_0 R_c^3$ gives (apart from a numerical constant) the number of prey within a spherical capture region and is a useful quantity for normalizations.

We found [10] an approximate simple empirical relation

$$J_e = C_A \frac{\epsilon^{1/3} D_0 R_c^3}{R_c^{2/3} + \eta_0^{2/3}}, \qquad (4)$$

which contains the parameter variation of the two limiting cases $R_c \ll \eta_0$ and $R_c \gg \eta_0$, as well as the correct cross-over length scale η_0. The effect of viscosity enters solely through η_0 in (4). Empirically, a coefficient $C_A \approx 7$ was found to give the numerical factors of the J_e-variations in the two relevant subranges correct within 15%. This result is consistent with numerical [10] as well as laboratory results [8]. The normalized expression $J_e/(D_0 R_c^3)$ based on (4) which will be used later is ill posed in the limit of $R_c \to 0$, with a diverging derivative there. This singularity has no physical consequences. Even for relatively large radii, e.g., $R_c = 10\eta$, we have contributions from viscosity, emphasizing that viscous effects can be more important than generally accepted for the present context. In Figure 4 we show (4) for varying ϵ and R_c in physical units taking ranges relevant for our database. We choose to normalize J_e with $D_0 R_c^3$ and a time scale t_m that will be defined later. This normalization is natural here, but gives rise to a "hidden" dependence on R_c^3.

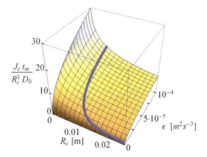

Figure 4. Illustration of normalized encounter rate $J_e t_m / (R_c^3 D_0)$ as a function of ϵ and R_c as obtained by (4). A blue line gives $R_c = \eta_0$, showing the separation between encounters with R_c in the inertial and viscous subranges, respectively. We took $t_m = 3$ s here.

The assumption of a spherical encounter surface is usually considered to be an oversimplification. More realistic models take a conical surface [32–34] with some opening angle θ, where hemispherical surfaces have often been assumed. There is empirical evidence that the results can be modified simply by introducing a multiplier for J_e in terms of a correction factor $\chi(\theta)$ which has an empirical form [29] where $\chi \approx 1$ for $\theta > 100°$ while

$$\chi(\theta) \approx 0.31\theta + 0.19\theta^2 - 0.06\theta^3, \quad (5)$$

with θ given in radians [29] for $0 \leq \theta < \pi/2$. The difference between $\chi(\theta)$ obtained for the viscous and inertial subranges is moderate [10], but in principle the ranges should be considered separately also in this respect.

The results quoted so far assumed both predators and prey to be passively following the turbulent motions in the flow. Self-induced motion patterns can also be taken into account by phenomenological models. For a simple cruising predator motion with prey passively following the flow, evidence was found [12] for a simple encounter rate expression in the form

$$J_e = \sqrt{J_0^2 + \left(D_0 \pi R_c^2 V_c h(\theta)\right)^2}, \quad (6)$$

where V_c is the cruising velocity, assumed constant, while the factor accounting for the "clearing surface" is $h(\theta) = \sin^2 \theta$ for $\theta < 90°$ and $h(\theta) = 1$ for $\theta > 90°$. The notation J_0 indicates the turbulence-induced encounter rate experienced by a predator at rest, with a seemingly universally useful approximation given by (4). The cruising velocity is measured in the local fluid element, and the root-mean-square (RMS) fluid velocity in the Eulerian frame of reference is not relevant for comparison; the Kolmogorov velocity u_K is more appropriate, see Table 1. It turns out that (6) can be used as an approximation for other motion patterns as well [12]. We note the similarity of (6) with other earlier results [5,35]. Results for travel-pause predators have been obtained also [12].

Table 1. Summary table of field data and some quantities derived from them.

Station	$\langle L_\ell \rangle$ [mm]	$\langle N_H \rangle$	D_0 [l^{-1}]	$\langle a \rangle$ [μm]	ϵ [m^{2}s^{-3}]	η_0 [mm]	τ_K [s]	u_K [m s^{-1}]
I	4.69	1.99	11.22	299	1.12×10^{-6}	17.11	1.16	1.14×10^{-3}
II	4.80	1.45	9.17	299	3.62×10^{-6}	12.77	0.64	1.53×10^{-3}
IIIA	4.72	2.09	19.74	245	5.44×10^{-5}	6.49	0.17	3.01×10^{-3}
IIIB	4.70	1.62	13.59	306	1.70×10^{-5}	8.68	0.30	2.24×10^{-3}
IV	4.22	3.34	7.90	214	4.25×10^{-8}	38.80	5.94	5.03×10^{-4}
V	4.69	1.26	3.25	263	2.46×10^{-6}	14.06	0.78	1.39×10^{-3}

The effects of a cruising velocity is illustrated in Figure 5a by showing (6) for $V_c = 10^{-3}$ m s^{-1} with its value for $V_c = 0$ subtracted. Figure 5b shows similar results for a 5 times larger cruising velocity. Evidently, the effects of self-induced cruising motions are strongest for small values of ϵ and small R_c. With the given parameters we find self-induced motions to have some significance for parts of the viscous subrange only.

Pioneering studies [5,6] discussed the encounter rate for ambush predators, but then included also the effects of self-induced motions. Only the inertial subrange of turbulence was considered in these early works. The capture success was assumed to be independent of the turbulence level. The importance of also the viscous subrange was realized later [10,29], see also a review [30]. This subrange turns out to be the most important one for the present study.

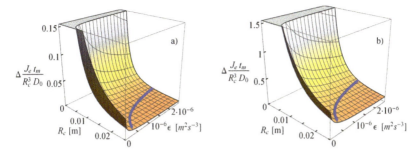

Figure 5. Illustration of the difference of the normalized encounter rate $J_e t_m/(D_0 R_c^3)$ for the case with and without a constant cruising velocity. We used $V_c = 10^{-3}$ m s^{-1} in a) and $V_c = 5 \times 10^{-3}$ m s^{-1} in (b). We took $t_m = 3$ s and a hemispherical field of view, $\theta = \pi/2$. A blue line indicates the separation line for R_c being in the viscous or the inertial subrange, as in Figure 4. Note the restricted range of ϵ as compared to Figure 4 and also the change in vertical scale in (a,b).

The analysis summarized in this section ignores intermittency effects in fluid turbulence. These account for the inhomogeneous distribution of the turbulent energy dissipation (at times called "the patchiness of turbulence"), for example manifested by spatially distributed interacting vortices. For individual realizations these can be observed to have consequences for organisms present in the flow [36–38]. The present study is concerned with phenomena occurring on the digestion time-scale which is of the order of 30–60 min. The intermittency effects will be smoothed out over these long times.

2.1.2. Capture Probability in Turbulent Environments

Several models can be found for the influence of turbulence on the capture rate of prey, given an encounter. The models depend on the species, predators as well as prey, and conditions in the surroundings, where turbulence is only one of the important parameters entering.

Reliable models for the encounter rates J_e of predators and prey in a turbulent environment can be found, at least for idealized models. We note, however, that the prey flux μ entering the gut is a more complex process. It is found to be an advantage to separate μ into an encounter rate, or clearance rate, J_e, and a capture probability P_c, give the encounter. Assuming the two to be statistically independent we have

$$\mu = J_e P_c. \tag{7}$$

Given independence of encounter and capture, the capture probability can be discussed independently of J_e. Several studies [13,14,39] assumed that P_c depends primarily on the time available for capture, other conditions considered constant. The simplest and often used model assumes capture with certainty ($P_c = 1$) if the time τ available is more that some characteristic time t_m, while $P_c = 0$ if the available time is $\tau < t_m$. To account for P_c we thus need the probability density $P_\tau(\tau)$ of times available for capture. This functional form has been determined empirically by numerical simulations of a turbulent flow [39] and presented for selected values of θ obtained in a form of a series approximation with tabulated coefficients. Other studies [13] used a simplified flow model where the problem could be solved analytically. For given species the time constant t_m depends on many conditions in the environment, light conditions [40] etc., but in particular on the age of the predators. By selecting a certain stage in development of fish larvae when establishing a database we can minimize a variation in t_m and assume it to be a constant. For the present study we selected cod larvae in the start-feeding phase (stage 7 larvae [21]). At this stage, the larvae have not yet developed a swim-bladder so any change in vertical motion apart from that induced by the initial and constant buoyancy requires swimming.

The main conclusion found by use of (7) can be summarized as: Turbulence is advantageous for predators by increasing the contact rate with prey as accounted for by J_e. Too large turbulence levels can however enhance the relative motions between predator and prey to an extent where the capture is strongly reduced because the time available for capture is too small [13,29,41–43]. This feature is accounted for by P_c. Due to a competition between the enhanced encounter rate and the reduced capture probability, there will be an optimum level of turbulence for predation. The prey capture rate as a function of the turbulence level will have a "dome shape", with a maximum at some intermediate turbulence level [13]. The optimum will vary among species.

Taking R_c to be in the viscous subrange of the turbulence, a compact form for the flux into the gut of ambush predators can be written [29] as

$$\mu = C_v D_0 R_c^3 \left(\frac{\epsilon}{\nu}\right)^{1/2} \chi(\theta) P_{es} \int_{t_m\sqrt{\epsilon/\nu}}^{\infty} P_{\tau'}(\tau')d\tau'. \tag{8}$$

A similar expression can be found for the inertial subrange. The escape of prey by their self-induced motion is assumed to be a statistically independent process and is here included by an empirical multiplier $0 < P_{es} \leq 1$ which has to be determined in a laboratory, for instance. Individual escape processes have been studied in detail [30]. Capture success rates in larval cod for quiet and weakly turbulent conditions have been studied in particular [19]. The integral in (8) accounts for the variation of the capture probability with the parameters of the problem. In (8) we recognize two length scales, the range of interception R_c and the average prey separation $D_0^{-1/3}$, with the product $D_0 R_c^3$ entering as a dimensionless parameter for the problem. We have μ being linearly proportional to the prey concentration D_0. The probability of two prey simultaneously entering the range of interception is assumed negligible. This implies that a predator can focus on one sample of prey at a time. Analytical approximations and tables of the probability density $P_\tau(\tau)$ needed in (8) can be found in the literature for various forms of the encounter and capture volumes [29]. The integral contribution to the results in (8) are given in terms of a normalized or dimensionless time $t_m\sqrt{\epsilon/\nu}$. Given the input parameters D_0 and ϵ with ν being a constant assumed to be known, we are thus in a position to give estimates for the average gut content of fish larvae when the organisms are characterized by their capture range R_c and opening angle θ for their field of view. An even more ambitious result is an estimate for the entire probability density of prey in the gut.

It has been suggested [29,44] that enhanced turbulence levels can be seen as "noise" that will make it difficult for a predator to discriminate signals from prey by disturbing the hydro-mechanical signals detected by the predators [45]. As a "rule of thumb" supported by analysis [44] we argue that if $10\tau_K \leq t_m$ we can expect that the turbulence induced noise-signal experienced by a predator will be disturbing and partially masking the flow disturbance induced by moving prey. For Station IIIB (see Table 1 and Figure 6) this can marginally be the case, but for the other stations this effect will have minor consequences and it is thus not included in the analysis.

2.1.3. Probability Density of Gut Content

Encountered and captured prey contributes to the gut content of the predator, and will remain detectable until it is digested after some time ("gut clearance rate"), here denoted τ_d. A simple model was proposed [18] for the gut content in form of a time series containing integers, $N = 1, 2, \ldots, N_m$ where N is the number of prey in the gut with N_m being some maximum gut capacity. The maximum gut capacity has only little consequence for the data presented in the following, but the general results can be useful for other data. A relatively simple analysis [18] including a finite gut capacity gave the result

$$P_d(N|N_m) = \frac{\exp(-\mu\tau_d)\Gamma(2+N_m)}{(1+N_m)\Gamma(1+N_m;\mu\tau_d)} \sum_{j=0}^{N_m} \delta_{j,N} \frac{(\mu\tau_d)^j}{j!} \tag{9}$$

for the normalized probability density in terms of the one-variable $\Gamma(y)$-function and the two-variables incomplete $\Gamma(y;z)$-function [46]. We introduced the Kronecker $\delta_{j,N}$ which is unity if $j = N$, and zero otherwise. We assume N_m to be given. The result has then no free adjustable parameters since the product $\mu\tau_d$ consists of measurable quantities. In nature we often found guts to be nearly empty and in such cases the finite gut capacity is of no consequence for the probability density. We can then use a simpler expression

$$P_d(N) = \sum_{j=0}^{\infty} \delta_{j,N} \frac{(\mu\tau_d)^j}{j!} \exp(-\mu\tau_d). \tag{10}$$

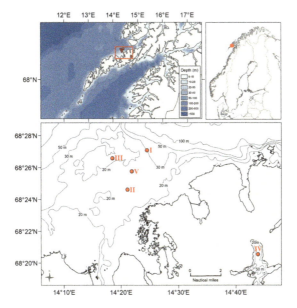

Figure 6. Site positions near Lofoten in Norway. The sites are marked by small filled red circles and roman numerals. Sites I, II, III, and V are in the open waters, while Site IV is deep in the Austnesfjord. At Site I the ocean depth is changing steeply from 20 m to 50 m, while Site III is positioned at a local plateau at a depth of 21 m. Sites II and V are at nearly the same depth, with similar variations in the surroundings, where Site II is closer to the shore. Additional details are reported elsewhere [18].

It is interesting to note that μ and τ_d appear in (9) as well as in (10), only as a product and not individually. From simple dimensional arguments [47] this could have been seen from the outset. We find this result to be important: if the distribution of gut content can be estimated and τ_d is known for the predator, we can find μ for the given organisms and conditions in the environment. This value for μ can then be compared to analytical results.

At first sight the arguments giving (9) and (10) seem to contain a flaw: it presumes N_m to be a fixed deterministic number and thereby all prey to have the same size. In reality we can have many small or a few large samples of prey in a full gut. This question was discussed elsewhere [18] with an assumed probability density for prey sizes. The result indicated that if the number of prey in the gut is large, some will be small, some large, so the actual net gut content would be close to the one obtained by assigning all prey the same size. In addition we note that the distribution of captured prey lengths (nauplii) had a distribution that was narrow compared to the average value, as demonstrated later.

2.2. Field Data

Our observational data are based on samples collected at Lofoten, Norway in April–May 1995 [18,48]. During this campaign data were collected at 5 sites, labeled Station I, II, III, IV and V, all located in shallow waters, see Figure 6. The conditions were strongly influenced by tidally-induced turbulence and turbulence induced from long swells. At one of the sites (Station III) the turbulence conditions changed during the day so we have two data-sets there, Station IIIA and Station IIIB. In the beginning of May the sun rises around 04 and settles around 22 local time. Sample collection was restricted to the time interval 06–22 in order to eliminate effects of reduced feeding/ingestion rates during the dark part of the day. The varying light conditions during the day were similar for each site [40].

The methodology of the plankton sampling and analysis, i.e. the vertical sampling of the cod larvae and their naupliar prey and the analysis of the gut content of the cod larvae, was adopted from previous studies [21,22]. The vertical concentration profile of fish larvae was determined by a fish larvae pump with a capacity of approximately 0.5–0.7 m^{3}s^{-1}. A plankton net with 375 µm mesh size was attached to the fish larvae pump. Samples were made at discrete depths from 5 m to 40 m. It was anticipated that the pump may cause some damage to the collected fish larvae. The damage is known to depend on the duration of the pumping times, in our case they varied from 15 s to 60 s. The quality of these data has therefore to be verified by another collection methods which do not have such a damaging effect. Fish larvae were also sampled by a vertically hauled plankton net. The opening of the net was 0.5 m^2 with a 375 µm mesh size. The net was hauled vertically from 50 m depth, or from approximately 2 m above the seabed when the water was shallower than 50 m. From the collected fish larvae those in the start-feeding phase (stage 7 larvae [21]) were selected and their lengths measured. Their gut content was inspected by counting the number of prey carcasses in the gut of fish larvae. The length distribution of cod larvae from a selected site is illustrated in Figure 7. Similar distributions were obtained from the other sites as well.

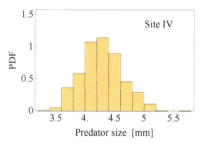

Figure 7. Summary figure showing the observed length distribution of cod larvae collected at Site IV using all data available for the site. The figure is thus based on 647 entries from the database. The scatter in the lengths is small compared to the average value $\langle L_\ell \rangle = 4.22 \pm 0.35$ due to the selection of the larvae. These figures are representative also for the subset of data obtained by use of the fine meshed net.

A measurement based on 299 cod larvae gave a distribution of the lengths a of the nauplii found in the guts, see Figure 8. The most probable nauplii length was found to be approximately 200 µm. The spread in the sizes is noticeably smaller than the average.

Figure 9 shows the vertical distribution of predators (here *Gadus morhua* L.) at Sites I to V. Data were collected at times evenly distributed over the day, but the number of samples collected varied from site to site. We find indications of a systematic variation with depth only at Site V and even here it is not significant.

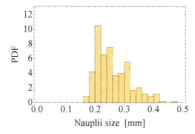

Figure 8. Probability density estimate for the distribution of the size a of the nauplii in the guts of cod larvae. The figure is based on the gut content of 299 cod larvae distributed over all 5 sites. The average value is $\langle a \rangle = 0.251 \pm 0.058$. The figure is taken from [18].

Figure 10 shows the vertical distribution of prey (here nauplii) at Sites I to V, sampled with a pump with plankton net mesh size 90 micrometers. Data were collected at times evenly distributed over the day, but the number of samples collected varied from site to site. We find no systematic variation of the density of nauplii with depth, and assume that the turbulent mixing is sufficient to smooth out gradients in the nauplii distribution for the present shallow waters. We note that the variability of measured nauplii concentration from one measurement to the other is the smallest for the site with smallest turbulence intensity. In the lack of any systematic depth variation of the nauplii distribution we assign each site as reference concentration the average over all depths for that site.

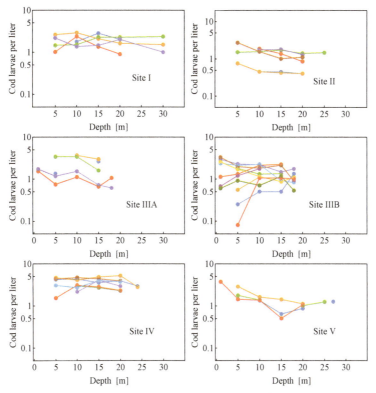

Figure 9. Vertical distribution of cod larvae at Sites I to V typically with 5 m resolution. In a few cases also some intermediate depths were sampled. The present data were obtained by the fish larvae pump as described elsewhere [18].

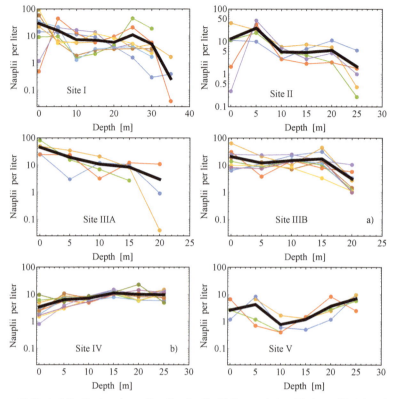

Figure 10. Vertical distribution of nauplii at Sites I to V with 5 m resolution. The heavy black line shows the average. Most often we find a small decrease in nauplii concentration with increasing depths.

Figure 11 shows scatter diagrams for the sizes of cod larvae and their gut contents. The correlations are in general not sufficient to allow conclusions of substance. The scatter in larvae size is moderate due to the pre-selection, and a high correlation between the size of cod larvae and their gut content can not be supported by the present database. Mostly we find a reasonable result indicating that large fish-larvae are most efficient in capturing prey.

The largest number of nauplii found in a gut was 12, and this number was observed only once. A gut content of 10 and 11 was both seen 4 times, while 9 nauplii were observed frequently. In the following we use $N_m = 9$ for all the fish larvae. Based on data obtained by the vertically hauled plankton net we estimated the distributions of the gut contents as shown in Figure 12. The net gives the least damage to the cod larvae, and the data for these are therefore analyzed separately. Filled circles in each of the Figure 12 give results derived by the analytical model (9) by adjusting the parameter $\mu\tau_d$ so that the average corresponds to the observed value of $\langle N \rangle$. The results were obtained for $N_m = 9$ as mentioned. Taking $N_m = 10$ gave modification that were noticeable only for Site IV, with results shown elsewhere [18]. In order to quantify the difference between the model results and the observations we note that for small $\mu\tau_d$, the model predicts $(\langle N^2 \rangle - \langle N \rangle^2)/\langle N \rangle = 1$. When taken as an average for all the datasets shown in Figure 12, the same quantity was found to be 1.18. We find this agreement to be sufficiently convincing to allow the model (9) to be used also more generally. For completeness we included with thin dashed lines in Figure 12 also the results found by using data collected via the fish larvae pump. Implicit in the derivation is the assumption that the gut content PDF's are time stationary.

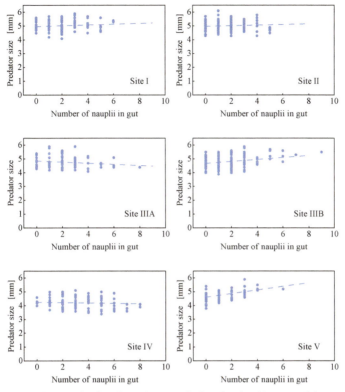

Figure 11. Scatter diagram for the relation between the lengths of cod-larvae and their gut content. The dashed lines are the best linear fits. The correlations are small, of the order of $R \approx 0.1$. In two cases the slope of the lines is negative but with a very small derivative. The largest correlation is found for Site V. Only the data obtained by the slowly moving net were used for the present figures. A projection of the figures on the vertical axis will give a result similar to the one shown in Figure 7.

Comparing the data found by the vertically hauled ichthyoplankton net and fish larvae pump as shown in Figure 12 we note a systematic overpopulation of empty guts (best seen for Site IIIA) in the data obtained by use of the pump. We take this as evidence that the pump is damaging plankton by making a significant part of them regurgitate. Existing and future data obtained by pumps like these should be interpreted with this possibility in mind. Because of the uncertainty associated with the data collected by the pump, we use only results for the vertically hauled net for comparison with analytical results. The depth resolution of the distribution of cod larvae and the corresponding gut content is then lost. The depth distribution of the number of cod larvae as found by the pump remains useful. Also earlier investigations [21,22] used the pump profiles to get an estimate of the vertical distributions of the larvae, while the gut contents were examined from the more gentle sampling by vertical net hauls.

Since the typical gut evacuation rates in first-feeding cod larvae are of the order of 1 h, it cannot be expected that the ambient food concentration of the cod larvae is equal to the sampling depth. Cod larvae are able to move vertically considerable distances during 1 h, up to 20 m. In addition, stronger mixed-layer turbulence can move the larvae even more during this period of time. Referring to Figure 10 we note that the vertical prey (nauplii) distribution was on average fairly uniform.

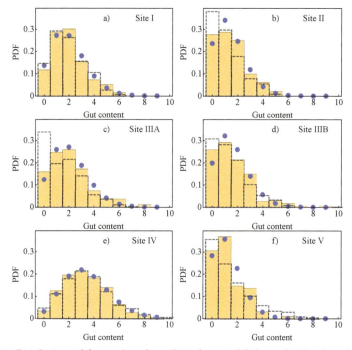

Figure 12. Distributions of the number of nauplii in the gut of *Gadus morhua* L. where the larvae were collected by a fine meshed net moved slowly from the seabed to the surface at Sites I through V (**a–f**). Corresponding analytical results are given by filled circles assuming $N_m = 9$. We have $\mu\tau_d = 1.99, 1.45, 2.09, 1.62, 3.34$, and 1.26. Thin dashed lines give the results where the data were collected by the pump "HUFSA". Parts of the figure are adapted from [18].

For a given average value of the capture rate μ for a population, some cod larvae will have more, some less than the number needed for survival. To give a convenient estimate for the gut contents we take the cumulative distributions for the number of prey in the guts of fish larvae (9) as shown in Figure 13 for the parameter values $\mu\tau_d = 1, 2, 3$ and 4. Taking for instance the case with $\mu\tau_d = 2$ (green curve) we find that a fraction 0.85 of the cod larvae have 3 nauplii or less in their gut, while for $\mu\tau_d = 4$ (red curve) this fraction is 0.45. When $\tau_d = 1$ h the values for μ are also the number of nauplii captured per hour. By the Figures 12 we found $\mu\tau_d = 1.99, 1.45, 2.09, 1.62, 3.34$, and 1.26. Taking here for illustration $\tau_d = 45$ min, we consequently have $\mu = 2.65, 1.93, 2.79, 2.16, 4.45$ and 1.68 captured nauplii per hour on average. These numbers should be compared to 2–3 nauplii per hour needed by cod larvae for survival.

In Table 1 we present a summary of averaged data as they are used for the comparison with analytical results. The set of observations at a given site are considered as individual realizations belonging to an ensemble with the given macroscopic parameters. The average number $\langle N_H \rangle$ of nauplii in gut is obtained by the reduced database found by using the fine meshed net moved slowly from the sea-bottom to the surface. The cod larvae mean length is $\langle L_\ell \rangle$, the concentration of nauplii in the surroundings is given by D_0, and the mean length of nauplii is $\langle a \rangle$. The specific turbulent energy dissipation is ϵ, and the derived effective Kolmogorov length η_0 is determined for each site as discussed in the following Section 2.2.1. The Kolmogorov time scale is $\tau_K = \sqrt{\nu/\epsilon}$.

Ambient temperatures ranged from 4.5° to 5.5 °C. The salinity ranged from 33.5 to 34, corresponding to a kinematic viscosity of the water $\nu = 1.5$ mm^{2}s^{-1} [49,50].

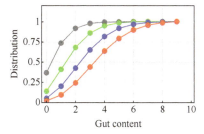

Figure 13. Cumulative distributions for the number of prey in the guts of fish larvae as determined by (9), here shown for the parameter values $\mu\tau_d = 1, 2, 3$ and 4 with grey, green, blue and red filled circles. These numbers cover the range of observed values in Figure 12. We took $N_m = 9$.

2.2.1. Estimates of the Average Specific Turbulent Energy Dissipation Rate

As we have seen, the turbulent energy dissipation rate ϵ is an essential parameter for describing turbulence in the inertial as well as the viscous subranges. When only a one-point measurement is available, the almost universally adopted method for determining ϵ relies on the Taylor hypothesis, or the frozen turbulence approximation [51–57]. It is there argued that a detected frequency spectrum can be "translated" to a wavenumber spectrum by $k \approx \omega/V$ where V is a constant average relative flow velocity between the rest frame of the fluid and the detector. In effect, it is assumed that the Doppler shift kV dominates the rest frame frequency ω', i.e., $\omega = \omega' + kV \approx kV$. Local homogeneity and isotropy of the turbulence is implicitly assumed when applying Taylor's hypothesis for interpreting turbulent power spectra. In case there is no average flow it will still be so that the small eddies in the inertial subrange are advected by the larges energy containing eddies [53] and the hypothesis can be applied also then, now with a suitably defined RMS fluid advection velocity. Details of the analysis and procedures used in the present work are reported elsewhere [55].

Fluctuating flow velocity components were detected by three different high resolution acoustic current meters (an Ocean ADV from NORTEK, a MINILAB and an UCM from SimTronix) were mounted on a submarine tower 6.5 m above the seabed [18,48]. All instruments were facing upwards in order to minimize possible effects of the construction on the observations. All data reported in the present work are obtained by the UCM, which measures the 3 velocity components of the fluctuating flow with a minimum resolvable wavelength of approximately 2 cm. In Figure 14 we show typical values for average horizontal velocities found to be in the range 5–10 cm s^{-1}. Most of the kinetic energy in the fluid motion is associated with the largest non-universal eddy dynamics. Taylor's hypothesis gives most accurate results for large velocities [52], but there seems to be no universally accepted reference velocity expressed in terms of the flow parameters.

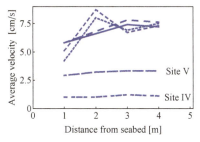

Figure 14. The average flow velocities are measured at the submerged tower at 4 levels above bottom. Full line gives data from Site I, long dashes for Site II, and shorter dashes for Sites IIIA and IIIB. The smallest value of ϵ is found at Site IV where also the average velocity is smallest. The figure is taken from [18].

Results for the experimentally obtained spectra for the three fluctuating fluid velocity components are shown in Figure 15. Due to the closeness of sea-bottom, the vertical component has a reduced intensity at small frequencies (corresponding to long wavelengths) but the three spectra are close for high frequencies (i.e., short wavelengths), where we argue that local isotropy and homogeneity has been reached for the small scales. The observation that a frequency spectrum with the Kolmogorov-Obukhov exponent at high frequencies, i.e., $\omega^{-5/3}$, is a good approximation can be taken as a support for the applicability of Taylor's hypothesis. The swells could be identified directly in the raw data as large amplitude intermittent oscillatory "bursts", and these contribute to the low frequency non-universal parts of the spectra.

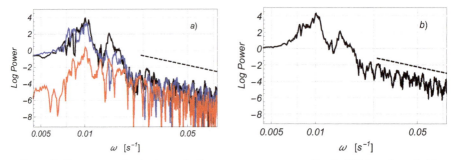

Figure 15. Power spectra for the three velocity components as measured by the UCM are shown in (**a**), with horizontal components in blue and black, vertical in red. In (**b**) we show the full power spectrum. Dashed reference lines have a slope of −5/3 as appropriate for the Kolmogorov-Obukhov spectrum. The raw spectral data have been smoothed by "binning" in the present representation. The figure is based on data from [18].

Power spectra for the three components of velocity fluctuations were measured at each site. The variations in the local flow velocity were sampled with a frequency of 2 Hz, using time series of 20 min duration. The average turbulence energy dissipation rate ϵ was determined by fitting experimentally obtained power spectra to the analytical Kolmogorov-Obukhov wave-number spectrum $C_{KO}\epsilon^{2/3}k^{-5/3}$ that contains ϵ. The universal Kolmogorov constant is $C_{KO} \approx 0.5$–1.5. As mentioned before, the comparison between the experimental frequency spectra and the theoretical wavenumber spectrum is made by reference to Taylor's hypothesis [51–53,55,58]. The robustness of the analysis giving ϵ is tested by using slightly different values of the exponent in the power-law, e.g., ω^{-2}.

Since the experimentally obtained spectral index agrees with the analytical −5/3-law, we have the main uncertainty in the estimate of ϵ to be in the use of Taylor's hypothesis and the uncertainty of the translational velocity being used. The experimentally obtained values of ϵ vary over the time series as can be seen in the relative variation $(\epsilon - \langle\epsilon\rangle)/\langle\epsilon\rangle$ at Site II, see Figure 16, where $\langle\epsilon\rangle$ is here the average value of ϵ for the given site. The Figure 16 is representative for the other sites as well [18]. Also the results for the turbulent energy dissipation ϵ are included in Table 1.

For conditions in the present study, the turbulence was dominated by tides and swells at the observation sites, and thus different from earlier studies [21] and [22] where wind-induced turbulence dominated.

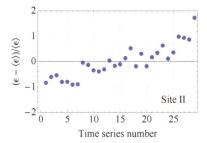

Figure 16. The relative scatter of ϵ over the different time-series as obtained at Site II.

2.2.2. Comparison of the Analytical Model for Capture Rates and the Field Data Results

We made a comparison between our theoretical model and data obtained by analyzing the field observations. We found that under statistically time stationary conditions the average value of the gut content gives an estimate of $\mu\tau_d$. Given an estimate for the digestion time τ_d, which has to be found by other means, in a laboratory for instance, we can then determine μ for the conditions specifying the predator and its environment. Gut evacuation rates were reported in [59], for instance. These rates can depend on temperature as well as other conditions [40]. These conditions can be assumed to be the same at all sites for our data. With the given selection of cod larvae we can assume that all predators are characterized by approximately the same parameters, so the differences are to be found in the environment. We have these parameters determined by the turbulence conditions ϵ and availability of prey D_0, both parameters measured, while the viscosity ν is determined by the temperature and salinity of the ambient water.

The comparison between the analysis and the present field data is shown in Figure 17. In previous studies of the dataset [18] it was noted that all predator sizes were below η_0 and thus in the viscous subrange, see also Table 1. The capture range R_c can however be larger than the cod larvae and it is then possible that the overlap region between viscous and inertial subranges have to be included in a complete analysis. This has a price, however: even after normalizing the capture rate by $D_0 R_c^3$ it is found that the complete analytical model depends explicitly on both ϵ and R_c, see for instance (4). The viscous subrange taken alone is simpler in this respect. In Figure 17 we thus show a surface spanned by the variables ϵ and R_c both in physical units. The variation of the capture rate is given through a normalization as $\mu t_m / (D_0 R_c^3)$. We assumed the minimum time needed for capture to be $t_m = 3$ s as a representative value [13]. Because of the uncertainties associated with the gut-content data obtained by the pump, we used only data from the vertically hauled plankton net for Figure 17. For the digestion time we took $\tau_d = 45$ min, although also longer times have been argued [59]. The analysis is trivially remedied to account for other values of τ_d, see for instance [18]. For the capture range we took $R_c = 2L_\ell$, see Table 1. This gives R_c in the range of 8–9 mm to be measured from the "center of mass" for the cod larvae. For the highest turbulence level, at Site IIIA, this value places R_c between the viscous and inertial subranges. This observation demonstrates the need for a model that encompasses both the inertial and the viscous subranges in describing the encounter and capture rates for predators and prey.

As an alternative presentation we show in Figure 18 a projection of the data-points on a plane that cuts the surface in Figure 17 at the position defined by the average of the 6 values of R_c's used there. The cut is shown by the blue line in Figure 18. Since the scatter in R_c-values as shown in Table 1 is moderate, we believe this figure also gives a useful illustration. A comparison with the analysis given elsewhere [18] illustrates the importance of including the correction from the inertial subrange. If this contribution is ignored, the dependence on R_c will as mentioned before only enter through the normalization of the vertical axis. In that case a 2 dimensional presentation as in Figure 18 will be adequate without the need of a projection. For R_c in the viscous subrange, the optimum capture value is found at $t_m \sqrt{\epsilon/\nu} \sim 1$ with $\mu \sim D_0 R_c^3/t_m$. The optimum thus obtained should be compared to the value of 2–3 nauplii captured per hour generally assumed to be needed for survival of cod larvae at the given stage. The digestion time τ_d does not enter here.

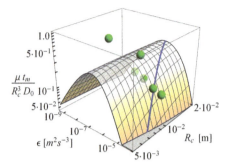

Figure 17. Comparison between analyzed field data (shown by small green spheres) and the theoretical model including the effects of turbulence on the encounter and capture rates. We assumed a hemispherical capture volume, $\theta = \pi/2$. Note that two data points are below the surface. See also results in [18]. The figure uses logarithmic scales on all axes. The variation in contact distances R_c is limited due to the pre-selection of cod larvae. A blue line separates viscous and inertial subranges as in Figure 4.

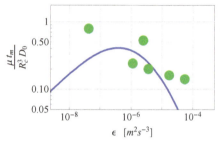

Figure 18. Double logarithmic presentation of a comparison between analyzed field data (shown by green filled circles) and the analytical model including the effects of turbulence on the encounter and capture rates. We assumed a hemispherical capture volume, $\theta = \pi/2$. See also results in [18]. The figure here is obtained by a projection on a plane that cuts the surface Figure 17 at the average of the 6 values of R_c. The sequence of the datapoints follow that of ϵ, i.e., from left to right we have the Sites IV, I, V, II, IIIB and IIIA.

The analysis and the present dataset supports, in particular, elements of a "dome shaped" capture probability [13] in the sense that we find a decreasing trend in the average capture probability for increasing large turbulence levels.

2.2.3. Consequences of Finite Sizes

As most other relates studies, also we described plankton as point-particles moving in the flow. A generalization of the study can be based on the Basset-Boussinesq-Oseen (BBO) equation derived for describing motions of finite size spherical particles in Stokes flows. A rigorous derivation [60] includes the effects of nonuniform velocity fields. The result is an analytical expression describing the motion of a spherical particle with given radius and mass, including the effects of gravitational acceleration. Most relevant aquatic organisms have mass density close to that of water. We therefore believe the finite size effects to be the most important, and will here briefly discuss some analytical models that can account for finite particle sizes [10,61]. These models represent a generalization of the Faxen corrections. The Stokes drag will be ignored here, although this effect can be included as well [61]. Our basic assumption is that a body with fixed shape responds to the moving flow by averaging the unperturbed space-time varying fluid velocity over its volume. This assumption seems reasonable and has shown promise when analyzing finite size particles moving in inhomogeneous pipe flows [62]. The averaging is most easily carried out in a spectral representation by a filtering, where the filter characteristics are determined by the particles (here plankton) in question [10,61]. For isotropic finite size particles we write an effective velocity as $v(\mathbf{r},t) = \int_{-\infty}^{\infty} \mathcal{F}(k)\mathbf{u}(\mathbf{k},t)\exp(i\mathbf{k}\cdot\mathbf{r})d\mathbf{k}$ where $\mathcal{F}(k)$ is a filter function. The only basic requirements are $\mathcal{F}(k \to 0) \to 1$ and $\mathcal{F}(k \to \infty) \to 0$. For spherical particles, the filtering is simple. Anisotropic particles have interesting flow properties [10,61], but these are of limited relevance here and will not be discussed.

To illustrate the consequences of finite particle sizes for the modeling we consider here spherical encounter surfaces with range R in the inertial subrange of turbulence, and will later discuss the problems in generalizing these results to the viscous subrange. Apart form a numerical constant, the relation (2) can be obtained by the longitudinal second order structure function

$$U^2(r) \equiv \left\langle \left(u_\|(\xi,t) - u_\|(\xi+r,t)\right)^2 \right\rangle,$$

written in one spatial dimension for simplicity, with more details given elsewhere [8]. We have the reference predator to be at position ξ and prey to be at $\xi + r$. For homogeneous isotropic turbulence conditions, we have $U(r)$ to be independent of ξ and only the separation r enters. For the inertial subrange we know that $U^2(r) \sim (\epsilon r)^{2/3}$. Taking the flux to a reference point particle (the predator in our case) through a spherical encounter surface with surface area πR^2 we can use the encounter estimate $J \sim D_0 U(R) R^2$, again apart from a numerical constant which is best determined empirically by experiments. Use of the previous expression for $U(R)$ in the inertial subrange gives $J \sim D_0 \epsilon^{1/3} R^{7/3}$ which is consistent with the result (2). With these results in mind we consider a finite size particle and apply the appropriate filtering to get a more general second order structure function in the form

$$U^2(r) = \left\langle \left(\int_{-\infty}^{\infty} \left(\mathcal{F}(k)u_\|(k,t) - u_\|(k,t)\exp(ikr)\right)\exp(ik\xi)dk\right)^2 \right\rangle, \qquad (11)$$

where we still consider prey as point particles, so that velocity is not filtered. After some manipulations reported elsewhere [63,64] an expression is obtained in the form

$$U^2(r) = \int_{-\infty}^{\infty} \mathcal{E}(k) \left(1 + |\mathcal{F}(k)|^2 - 2\mathcal{F}(k)\exp\left(-\frac{1}{2}k^2 r^2\right)\right) dk, \qquad (12)$$

where the power spectrum for the fluctuating velocity is $\mathcal{E}(k) \equiv \langle u_\|(k,t)u_\|(-k,t)\rangle = \langle |u_\|(k)|^2\rangle$ for time stationary conditions. At one place we approximated the average of a product by the product of two averages to have $\mathcal{E}(k)$ appearing explicitly and used also $\langle\exp(ikr)\rangle = \exp(-k^2 r^2/2)$ as for Gaussian processes. If we ignore intermittency corrections and take the Kolmogorov-Obukhov spectrum as $\mathcal{E}(k) \sim \epsilon^{2/3} k^{-5/3}$ in (12), together with the limiting form $\mathcal{F}(k) \approx 1$ for point particles we recover $J \sim D_0 U(R) R^2 \sim D_0 \epsilon^{1/3} R^{7/3}$ analytically for $r = R$. The other limiting case where $\mathcal{F}(k) \approx 0$ corresponds to a predator at rest, i.e., a "tethered" predator, where $J \sim D_0 R^2 \sqrt{\langle u^2 \rangle} \gg D_0 \epsilon^{1/3} R^{7/3}$ where $\langle u^2 \rangle = \int_{-\infty}^{\infty} \mathcal{E}(k) dk$. This latter result has experimental support [8]. For the case where R is much larger than the reference particle, we have $\exp(-k^2 R^2/2) \approx 0$ while $\mathcal{F}(k) \approx 1$; in this limit the finite size of the predator is immaterial, as also expected intuitively. The intermediate case with some general form for the filter $\mathcal{F}(k)$ is more problematic and will depend on the shape modeled by $\mathcal{F}(k)$.

For illustration we assume $\mathcal{F}(k) = \exp(-k^2 S^2)$ where S represents a characteristic size of the predator, $S < R$. Using again the Kolmogorov-Obukhov spectrum for $\mathcal{E}(k)$ we have

$$J \sim D_0 \epsilon^{1/3} R^2 \sqrt{(R^2 + 2S^2)^{1/3} - S^{2/3} 2^{-1/3}} < D_0 \epsilon^{1/3} R^{7/3} \quad \text{for} \quad R > S. \tag{13}$$

Within this model we find that the encounter rate is reduced compared to the results for plankton moving like point particles, i.e., $J_{e,I}$ in (2). Taking $R \approx 2S$ as used in the foregoing analysis we have a reduction of approximately 10% as compared to $J_{e,I}$.

We have thus found a viable model for including finite size effects where some limiting cases can be tested, provided we restrict the analysis to the inertial subrange of turbulence. This subrange is, however, restricted to wavenumbers smaller than the Kolmogorov wavenumber. The problem in generalizing the model to the viscous subrange is due to the lack of a universal spectral model for this limit. The analysis of Heisenberg [65] predicted a $\mathcal{E}(k) \sim k^{-7}$ power law spectral variation for large k, but this result makes $\int_{-\infty}^{\infty} k^{2p} \mathcal{E}(k) dk$ diverge for sufficiently large p-values, and the same can be argued for any power-law $\mathcal{E}(k) \sim k^{-m}$. This divergence is considered to be unphysical [66] and for the time being we have no generally accepted spectrum for the viscous subrange. For the second order structure function we have no similar problems. Assuming that the trends found for the inertial subrange apply to the viscous subrange as well, we argue that finite size effects give rise to a reduction factor of 10% also here. The uncertainty introduced by this is comparable to or even less than other uncertainties in the problem. Finite size effects are considered to have minor importance for our analysis but might be relevant other studies.

3. Discussion

We took the digestion time τ_d to be constant. In principle it is possible for τ_d to depend on the gut content. Our data give no support for such models [59]. Should that be the case, we would observe a systematic overpopulation for small or for large gut contents as compared to our model. It is possible that such a relation can be found by studies of populations where full guts are more frequent than in our database.

In the presentations in Figure 17 we ignored the prey escape reactions by setting $P_{es} = 1$ in (8). Since P_{es} accounts for the fraction of the prey density D_0 that escapes capture, we can change the normalization on the vertical axis in Figure 17 to $\mu t_m / (D_0 P_{es} R_c^3)$ and the analytical part of the figure will then apply more generally. The points for the observations (green spheres) will then have to be moved upwards for $P_{es} < 1$.

The assumed value for the minimum time needed for capture, $t_m = 3$ s, may appear to be large, but it represents an average of the minimum time required for capture and the time needed for capture with large probability [29]. With this in mind we find the value $t_m = 3$ s to be reasonable. The most sensitive parameter in the analytical model is R_c by entering to the power 3.

The largest deviation between the data and the analytical model is found for the smallest turbulence level at Site IV at the head of the Austnesfjord. We have no verifiable explanation for this deviation, but note that self-induced motions will have the strongest effects for small turbulence levels, see (6) and Figure 5. The present analysis assumes that the cod larvae are ambush predators and ignores self-induced motions entirely. This assumption need not be strictly fulfilled, and we see this as a possible explanation for the noted disagreement. Self-induced motions will enhance the encounter rate. Using (6) we find, however, that excessively large values of V_c need to be introduced in order to make the analytical result close to observations. Based on our data, we find however one conclusion to be inescapable: turbulence matters!

Possible sources for errors in the data analysis, be it systematic or random, have been discussed elsewhere [18] and the comments there apply equally well to the present analysis. We believe that the most significant uncertainty is associated with the estimates of ϵ, see also Figure 16.

The conditions characterizing the present set of field data differs from other studies by the source of the turbulence generation. With the exception of Site IV placed at the head of the Austnesfjord, we find the dominant energy sources to be swells and tidal motions, in contrast to wind generated turbulence [67]. Results for this latter case are usually presented in terms of an ocean surface wind velocity W and a depth Z, referring to a an empirical relation

$$\epsilon = 5.8 \times 10^{-9} \frac{W^3}{Z}, \tag{14}$$

expressing ϵ in units of m^{2}s^{-3}, W in m s^{-1} and Z in m. For such conditions a predator can change its vertical position to obtain optimum conditions [68,69]. For our dataset the model (14) is not applicable. Wind generated turbulence was the dominant source of turbulence in other studies [21,22].

The target species of the present study, i.e., first-feeding cod larvae preying on copepod nauplii, and the biological sampling methods, as outlined above, were the same as those used in the earlier field studies [21,22]. The study locations and the methods representing ambient turbulence for the organisms, however, differed substantially. While the studies [21,22] were conducted above the deep-water areas (bottom depths 80–150 m) of the Lofoten nursery grounds, where pelagic-layer turbulence is mainly produced by wind mixing, the present study was conducted in the shallow-water regions (bottom depths 17–30 m) where the turbulence is largely produced by near bottom tidal current shears and also by swells. The two earlier studies [21,22] were conducted without measuring mixed-layer turbulence, since the authors at that time did not have access to free-falling airfoil probes [70] to profile turbulence through the upper layer. Therefore, an empirical relationship established between winds and turbulent energy dissipation rates [23] was applied to estimate the ambient turbulence of the organisms studied. The present study, on the other hand, was motivated by how tidal-induced turbulence, generated from the bottom, might differ from wind-induced turbulence of the mixed layer, generated from the surface of the ocean. The turbulence measurements used here, were made with acoustic current meters on top of a 6.5 m high tower deployed on the seabed. Differently from profiling airfoil probes, this allowed for generating time series of turbulence measurements. On the other hand, measurements were here limited to one vertical position.

As outlined before, the biological sampling technique of measuring the stomach fullness of a fish larvae predator, here quantified as the number of nauplii in the gut of the predator, reflects the integrated encounters between predator and prey over a time period of about 1–2 h prior to sampling. During this time span the predator can migrate and/or become transported vertically through the mixed-layer. Moreover, the predators and the prey will typically become horizontally advected by the currents at a distance of 0.5–2 km [71] over the same period of time. This justifies application of vertically averaged prey concentrations when comparing the number of prey in the gut of the predator. However, the two other major factors influencing predator-prey encounter rates, namely the turbulence and light conditions can both change systematically through the pelagic layer. Firstly, the turbulent energy dissipation rate generated from surface wind mixing is not vertically homogeneous through the upper part of the mixing layer as suggested by [23] but decreases by some empirical function [13] as for example in (14). Recent empirical studies based on the measurements from state-of-the-art turbulence profiler show that the turbulent energy dissipation rate changes with depth as $Z^{-1.15}$ over the mixing layer [72]. Secondly, there are indications of a certain variation in encounter rate with light intensity [22]. Since this intensity changes with depth as $e^{-\kappa Z}$, where $\kappa \sim 0.14$ m^{-1} for coastal waters [16], it is also expected that the vertical position of the predator will affect the prey encounter rate.

In summary, of the three factors influencing encounter rates for larval predators of the wind-mixed layer, i.e., prey concentration, light intensity and turbulence, the two latter factors have the most substantial (and systematic) decrease with depth. This implies that the exact vertical position within the mixed layer will have large impacts on the encounter rates experienced by the predator in a wind mixed layer.

From field observations of spatial distributions of first-feeding cod larvae we know that their vertical distributions vary considerably with light conditions and with the wind mixing [73,74]. Generally, first-feeding cod larvae are found at highest concentrations between 10 and 30 m depth, but there are large short-time variations over a wider vertical range, typically from 5 to 40 m depth [73]. During stronger wind mixing, larvae are dispersed vertically deeper down to even 50 m depth by vertical mixing/and or active migration [73]. This implies that the cod larvae by a change in vertical position may be able to adjust their turbulence-induced encounter rates to the optimum [13] and thereby avoid turbulence levels exceeding those for optimum encounter-capture rates. Over the vertical range observed for cod larvae (i.e., surface layer to 50 m depth) the wind-generated turbulent energy dissipation rate changes by two orders of magnitude [72]. The previous field experiments [21,22] covered wind-induced turbulence for wind speeds up to 11 m s^{-1}. Most of the observations summarized there did not reach values of encounter rates for the larvae exceeding the optimal local maximum illustrated in, for instance, Figure 18. One single observation site in the former studies [22] was located in the same shallow-water and tidal-energetic region as for the present data. The Greenberg relation for turbulent energy dissipation rates [75] applied for those conditions indicated a turbulent energy dissipation rate ϵ of the order of 10^{-6} m^{2}s^{-3}. The observed larval encounter rate for this situation was higher than for the highest wind-induced turbulent conditions (11 m s^{-1}). The fish larvae were not exposed to turbulence levels exceeding the optimum of the "dome shaped" capture rate. Differently, in the present study of shallow-water depths of 17 to 30 m, the turbulent energy dissipation rates reached levels up to 10^{-4} m^{2}s^{-3}. With no possibility for the larvae to escape to greater depth, they became exposed to turbulence levels exceeding the optimum for encounters followed by capture.

4. Conclusions

The available data demonstrate the importance of the viscous subrange for describing the effects of turbulence on feeding conditions for plankton, but show also that a model is needed to account for the transition between viscous and inertial subranges of fully developed turbulence. Some of our data sampled in a shallow-water environment with high turbulence, demonstrate the reduction of capture probabilities caused by strong turbulent motions. Some cases are near the optimum level of the "dome shaped" capture rate. Only one case, from Site IV, falls in the range where turbulence has a purely advantageous effect. This site is located deep in a fjord with a lower level of turbulence. Within our database, the most advantageous site for the cod larvae is thus clearly in the Austnesfjord. Future studies of this problem should bear in mind that the relevant turbulence range is determined also by the contact range R_c of the predator, so a given turbulence level can appear strong for some organisms, and weak for others.

The ideas advanced in the present study referred explicitly to aquatic organisms in a turbulent environment. Elements of the models may, however, have wider applicability. In discussions with one of the authors (HLP), Prof. Jukka Heikkinen draws attention to a different problem concerning burning of charcoal dust. This process is relevant, for instance, for disposing of charcoal in filters. The burning is accelerated by dispersing the charcoal dust in a very hot air (i.e., flames). In this case we can associate the charcoal particles (until they burn out) with the predators, while their prey is the Oxygen in the surrounding. The flames constitute the turbulent environment giving the enhanced mixing. In this case we can also assume the turbulence conditions to be well described by the inertial range of the Kolmogorov-Obukhov law and also include the universal dissipation range. We anticipate that the analysis of the present study can be generalized to account also for elements in the problem outlined here being aware, though, that very different effects are also at play there, such as radiative and thermal effects, and the fact the dust particles may adhere to each other.

Author Contributions: Data acquisition and analysis, J.E.S. and S.S.; project administration, S.S.; original draft preparation, H.L.P., S.S. and J.K.T.; statistical analysis of data, J.E.S. and H.L.P.; theory and numerical simulations, H.L.P. and J.K.T. All authors have read and agreed to the published version of the manuscript.

Funding: Parts of the study were carried out under the project *"Feeding conditions of cod larvae in an area with high tidal energy"* funded with support from the Norwegian Research Council (NRC). We also received support from the ECOBE project.

Acknowledgments: The authors thank Jan Even Øie Nilsen for help with the turbulence measurements and Bjørnar Ellertsen together with Petter Fossum for assistance with the data collection and analysis. We thank also Karen Gjertsen for her assistance with producing Figure 6. Valuable discussions with Øyvind Fiksen are gratefully acknowledged.

Conflicts of Interest: The authors declare no conflict of interest.

References

1. Houde, E.D. Fish early life dynamics and recruitment variability. *Am. Fish. Soc. Symp.* **1987**, *2*, 17–29.
2. Houde, E.D. Subtleties and episodes in the early life of fishes. *J. Fish Biol.* **1989**, *35*, 29–38. [CrossRef]
3. Hinrichsen, H.H.; Möllmann, C.; Voss, R.; Köster, F.W.; Kornilovs, G. Biophysical modeling of larval Baltic cod (*Gadus morhua*) growth and survival. *Can. J. Fish. Aquat. Sci.* **2002**, *59*, 1858–1873. [CrossRef]
4. Lough, R.G.; Buckley, L.J.; Werner, F.E.; Quinlan, J.A.; Pehrson Edwards, K. A general biophysical model of larval cod (*Gadus morhua*) growth applied to populations on Georges Bank. *Fish. Oceanogr.* **2005**, *14*, 241–262. [CrossRef]
5. Rothschild, B.J.; Osborn, T.R. Small-scale turbulence and plankton contact rates. *J. Plankton Res.* **1988**, *10*, 465–474. [CrossRef]
6. Osborn, T. The role of turbulent diffusion for copepods with feeding currents. *J. Plankton Res.* **1996**, *18*, 185–195. [CrossRef]
7. Hill, P.S.; Nowell, A.R.M.; Jumars, P.A. Encounter rate by turbulent shear of particles similar in diameter to the Kolmogorov scale. *J. Mar. Res.* **1992**, *50*, 643–668. [CrossRef]
8. Mann, J.; Ott, S.; Pécseli, H.L.; Trulsen, J. Turbulent particle flux to a perfectly absorbing surface. *J. Fluid Mech.* **2005**, *534*, 1–21. [CrossRef]
9. Boffetta, G.; Pécseli, H.L.; Trulsen, J. Numerical studies of turbulent particle fluxes into perfectly absorbing spherical surfaces. *J. Turbul.* **2006**, *7*, N22. [CrossRef]
10. Pécseli, H.L.; Trulsen, J. Turbulent particle fluxes to perfectly absorbing surfaces: A numerical study. *J. Turbul.* **2007**, *8*, N42. [CrossRef]
11. Lewis, D.M.; Pedley, T.J. Planktonic contact rates in homogeneous isotropic turbulence: Theoretical predictions and kinematic simulations. *J. Theor. Biol.* **2000**, *205*, 377–408. [CrossRef] [PubMed]
12. Pécseli, H.L.; Trulsen, J.; Fiksen, Ø. Predator-prey encounter rates in turbulent water: Analytical models and numerical tests. *Prog. Oceanogr.* **2010**, *85*, 171–179. [CrossRef]
13. MacKenzie, B.R.; Miller, T.J.; Cyr, S.; Leggett, W.C. Evidence for a dome-shaped relationship between turbulence and larval fish ingestion rates. *Limnol. Oceanogr.* **1994**, *39*, 1790–1799. [CrossRef]
14. Jenkinson, I.R. A review of two recent predation-rate models: The dome-shaped relationship between feeding rate and shear rate appears universal. *ICES J. Mar. Sci.* **1995**, *52*, 605. [CrossRef]
15. MacKenzie, B.R.; Kiørboe, T. Larval fish feeding and turbulence: A case for the downside. *Limnol. Oceanogr.* **2000**, *45*, 1–10. [CrossRef]
16. Fiksen, Ø.; Utne, A.; Aksnes, D.; Eiane, K.; Helvik, J.; Sundby, S. Modelling the influence of light, turbulence and ontogeny on ingestion rates in larval cod and herring. *Fisheries Oceanog.* **1998**, *7*, 355–363. [CrossRef]
17. Vollset, K.; Folkvord, A.; Browman, H. Foraging behaviour of larval cod (*Gadus morhua*) at low light intensities. *Mar. Biol.* **2011**, *158*, 1125–1133. [CrossRef]
18. Pécseli, H.L.; Trulsen, J.K.; Stiansen, J.E.; Sundby, S.; Fossum, P. Feeding of plankton in turbulent oceans and lakes. *Limnol. Oceanogr.* **2019**, *64*, 1034–1046. [CrossRef]
19. Solberg, T.; Tilseth, S. Growth, energy consumption and prey density requirements in first feeding larvae of cod (*Gadus morhua* L.). In Proceedings of the Propagation of Cod *Gadus morhua* L., Arendal, Noway, 14–17 June 1983.
20. Sundby, S. Turbulence and ichthyoplankton: Influence on vertical distributions and encounter rates. *Sci. Mar.* **1997**, *61*, 159–176.
21. Sundby, S.; Fossum, P. Feeding conditions of arcto-Norwegian cod larvae compared with the Rothschild-Osborn theory on small-scale turbulence and plankton contact rates. *J. Plankton Res.* **1990**, *12*, 1153–1162. [CrossRef]

22. Sundby, S.; Ellertsen, B.; Fossum, P. Encounter rates between first-feeding cod larvae and their prey during moderate to strong turbulent mixing. *ICES Mar. Sci. Symp.* **1994**, pp. 393–405.
23. Oakey, N.S.; Elliott, J.A. Dissipation within the surface mixed layer. *J. Phys. Oceanogr.* **1982**, *12*, 171–185. [CrossRef]
24. Biferale, L.; Boffetta, G.; Celani, A.; Devenish, B.; Lanotte, A.; Toschi, F. Multifractal statistics of Lagrangian velocity and acceleration in turbulence. *Phys. Rev. Lett.* **2004**, *93*, 064502. [CrossRef] [PubMed]
25. Biferale, L.; Boffetta, G.; Celani, A.; Lanotte, A.; Toschi, F. Particle trapping in three-dimensional fully developed turbulence. *Phys. Fluids* **2005**, *17*, 021701. [CrossRef]
26. Tennekes, H.; Lumley, J.L. *A First Course in Turbulence*; The MIT Press: Cambridge, MA, USA, 1972.
27. Sreenivasan, K.R. On the universality of the Kolmogorov constant. *Phys. Fluids* **1995**, *7*. [CrossRef]
28. Davidson, P.A. *Turbulence. An Introduction for Scientists and Engineers*; Oxford University Press: Oxford, UK, 2004.
29. Pécseli, H.L.; Trulsen, J.; Fiksen, Ø. Predator-prey encounter and capture rates for plankton in turbulent environments. *Prog. Oceanogr.* **2012**, *101*, 14–32. [CrossRef]
30. Kiørboe, T. *A Mechanistic Approach to Plankton Ecology*; Princeton Univ. Press: Princeton, NJ, USA, 2008.
31. Pigolotti, S.; Jensen, M.H.; Vulpiani, A. Absorbing processes in Richardson diffusion: Analytical results. *Phys. Fluids* **2006**, *18*, 048104. [CrossRef]
32. Lewis, D.M.; Pedley, T.J. The influence of turbulence on plankton predation strategies. *J. Theor. Biol.* **2001**, *210*, 347–365. [CrossRef]
33. Lewis, D.M.; Bala, S.I. Plankton predation rates in turbulence: A study of the limitations imposed on a predator with a non-spherical field of sensory perception. *J. Theor. Biol.* **2006**, *242*, 44–61. [CrossRef]
34. Mann, J.; Ott, S.; Pécseli, H.L.; Trulsen, J. Laboratory studies of predator-prey encounters in turbulent environments: Effects of changes in orientation and field of view. *J. Plankton Res.* **2006**, *28*, 509–522. [CrossRef]
35. MacKenzie, B.R.; Kiørboe, T. Encounter rates and swimming behaviour of pause-travel and cruise larval fish predators in calm and turbulent laboratory environments. *Limnol. Oceonogr.* **1995**, *40*, 1278–1289. [CrossRef]
36. Reigada, R.; Hillary, R.M.; Bees, M.A.; Sancho, J.M.; Sagués, F. Plankton blooms induced by turbulent flows. *Proc. Royal Soc. London. Ser. B Biol. Sci.* **2003**, *270*, 875–880. [CrossRef] [PubMed]
37. Durham, W.M.; Climent, E.; Barry, M.; De Lillo, F.; Boffetta, G.; Cencini, M.; Stocker, R. Turbulence drives microscale patches of motile phytoplankton. *Nat. Comm.* **2013**, *4*, 2148. [CrossRef]
38. Breier, R.E.; Lalescu, C.C.; Waas, D.; Wilczek, M.; Mazza, M.G. Emergence of phytoplankton patchiness at small scales in mild turbulence. *Proc. Natl. Acad. Sci. USA* **2018**, *115*, 12112–12117. [CrossRef] [PubMed]
39. Pécseli, H.L.; Trulsen, J.K.; Fiksen, Ø. Predator-prey encounter and capture rates in turbulent environments. *Limnol. Oceanog. Fluids Environ.* **2014**, *4*, 85–105. [CrossRef]
40. Dam, H.G.; Peterson, W.T. The effect of temperature on the gut clearance rate constant of planktonic copepods. *J. Exp. Mar. Biol. Ecol.* **1988**, *123*, 1–14. [CrossRef]
41. Marrasé, C.; Costello, J.H.; Granata, T.; Strickler, J.R. Grazing in a turbulent environment: Energy dissipation, encounter rates, and efficacy of feeding currents in Centropages hamatus. *Proc. Nat. Acad. Sci. USA* **1990**, *87*, 1653–1657. [CrossRef] [PubMed]
42. Saiz, E.; Kiørboe, T. Predatory and suspension-feeding of the copepod *Acartia-tonsa* in turbulent environments. *Mar. Ecol. Prog. Ser.* **1995**, *122*, 147–158. [CrossRef]
43. Kiørboe, T.; Saiz, E. Planktivorous feeding in calm and turbulent environments, with emphasis on copepods. *Mar. Ecol. Prog. Ser.* **1995**, *122*, 135–145. [CrossRef]
44. Pécseli, H.L.; Trulsen, J.K. Plankton's perception of signals in a turbulent environment. *Adv. Phys. X* **2016**, *1*, 20–34. [CrossRef]
45. Kiørboe, T.; Visser, A.W. Predator and prey perception in copepods due to hydromechanical signals. *Marine Ecol. Prog. Ser.* **1999**, *179*, 81–95. [CrossRef]
46. Abramowitz, M.; Stegun, I.A. *Handbook of Mathematical Functions with Formulas, Graphs, and Mathematical Tables*; Dover: New York, NY, USA, 1972.
47. Buckingham, E. On physically similar systems; illustrations of the use of dimensional equations. *Phys. Rev.* **1914**, *4*, 345–376. [CrossRef]

48. Gytre, T.; Nilsen, J.E.O.; Stiansen, J.E.; Sundby, S. Resolving small scale turbulence with acoustic Doppler and acoustic travel time difference current meters from an underwater tower. In Proceedings of the OCEANS 96 MTS/IEEE Conference Proceedings, The Coastal Ocean—Prospects for the 21st Century, Fort Lauderdale, FL, USA, 23–26 September 1996; pp. 442–450.
49. Sharqawy, M.H.; Lienhard, J.H.; Zubair, S.M. Thermophysical properties of seawater: A review of existing correlations and data. *Desalin. Water Treat.* **2010**, *16*, 354–380. [CrossRef]
50. Sharqawy, M.H.; Lienhard, J.H.; Zubair, S.M. Erratum to Thermophysical properties of seawater: A review of existing correlations and data [Desalination and Water Treatment, Vol. 16 (2010) 354-380]. *Desalin. Water Treat.* **2012**, *44*, 361–361. [CrossRef]
51. Wandel, C.F.; Kofoed-Hansen, O. On the Eulerian-Lagrangian transformation in the statistical theory of turbulence. *J. Geophys. Res.* **1962**, *67*, 3089–3093. [CrossRef]
52. Shkarofsky, I.P. *Turbulence in Fluids and Plasmas*; Chapter "Analytic Forms for Decaying Space/Time Turbulence Functions"; Polytechnic Press: Brooklyn, NY, USA, 1969; pp. 289–301.
53. Tennekes, H. Eulerian and Lagrangian time microscales in isotropic turbulence. *J. Fluid Mech.* **1975**, *67*, 561–567. [CrossRef]
54. Wyngaard, J.C.; Clifford, S.F. Taylor's hypothesis and highfrequency turbulence spectra. *J. Atmos. Sci.* **1977**, *34*, 922–929. [CrossRef]
55. Stiansen, J.E.; Sundby, S. Improved methods for generating and estimating turbulence in tanks suitable for fish larvae experiments. *Sci. Mar.* **2001**, *65*, 151–167. [CrossRef]
56. Trujillo, J.J.; Trabucchi, D.; Bischoff, O.; Hofsäß, M.; Mann, J.; Mikkelsen, T.; Rettenmeier, A.; Schlipf, D.; Kühn, M. Testing of frozen turbulence hypothesis for wind turbine applications with a Staring Lidar. *Geophys. Res. Abstr.* **2010**, *12*, 5410.
57. Geng, C.; He, G.; Wang, Y.; Xu, C.; Lozano-Durán, A.; Wallace, J.M. Taylor's hypothesis in turbulent channel flow considered using a transport equation analysis. *Phys. Fluids* **2015**, *27*, 025111. [CrossRef]
58. Larsén, X.G.; Vincent, C.; Larsen, S. Spectral structure of mesoscale winds over the water. *Quart. J. Roy. Meteorol. Soc.* **2013**, *139*, 685–700. [CrossRef]
59. Tilseth, S.; Ellertsen, B. Food consumption rate and gut evacuation processes of first-feeding cod larvae (*Gadus Morhua* L.). In Proceedings of the Propagation of Cod *Gadus morhua* L., Arendal, Noway, 14–17 June 1983.
60. Maxey, M.R.; Riley, J.J. Equation of motion for a small rigid sphere in a nonuniform flow. *Phys. Fluids* **1983**, *26*, 883–889. [CrossRef]
61. Pécseli, H.L.; Trulsen, J. Predator-prey encounter rates in turbulent environments: consequences of inertia effects and finite sizes. In *From Leonardo to ITER: Nonlinear and Coherence Aspects, Proceedings of the AIP Conference*; American Institute of Physics: Melville, NY, USA, 2009; Volume 1177, pp. 85–95.
62. Batchelor, G.K.; Binnie, A.M.; Phillips, O.M. The mean velocity of discrete particles in turbulent flow in a pipe. *Proc. Roy. Soc. Lond.* **1955**, *B 68*, 1095–1104. [CrossRef]
63. Mikkelsen, T.; Larsen, S.E.; Pécseli, H.L. Diffusion of Gaussian puffs. *Quart. J. Roy. Meteorol. Soc.* **1987**, *113*, 81–105. [CrossRef]
64. Misguich, J.H.; Balescu, R.; Pécseli, H.L.; Mikkelsen, T.; Larsen, S.E.; Xiaoming, Q. Diffusion of charged particles in turbulent magnetoplasmas. *Plasma Phys. Contr. Fusion* **1987**, *29*, 825–856. [CrossRef]
65. Heisenberg, W. Zur statistische theorie der turbulenz. *Z. Phys.* **1948**, *124*, 628–657. [CrossRef]
66. Pécseli, H.L. *Low Frequency Waves and Turbulence in Magnetized Laboratory Plasmas and in the Ionosphere*; IOP Publishing: Bristol, UK, 2016. [CrossRef]
67. MacKenzie, B.R.; Leggett, W.C. Wind-based models for estimating the dissipation rates of turbulent energy in aquatic environments: empirical comparisons. *Mar. Ecol. Prog. Ser.* **1993**, *94*, 207–216. [CrossRef]
68. Maar, M.; Visser, A.W.; Nielsen, T.G.; Stips, A.; Saito, H. Turbulence and feeding behaviour affect the vertical distributions of *Oithona similis* and *Microsetella norwegica*. *Mar. Ecol. Prog. Ser.* **2006**, *313*, 157–172. [CrossRef]
69. Tanaka, M. Changes in vertical distribution of zooplankton under wind-induced turbulence. *Fluids* **2019**, *4*, 195. [CrossRef]
70. Oakey, N.S. Determination of the rate of dissipation of turbulent energy from simultaneous temperature and velocity shear microstructure measurements. *J. Phys. Oceanogr.* **1982**, *12*, 256–271. [CrossRef]

71. Strand, K.O.; Vikebø, F.; Sundby, S.; Sperrevik, A.K.; Breivik, Ø. Subsurface maxima in buoyant fish eggs indicate vertical velocity shear and spatially limited spawning grounds. *Limnol. Oceanog.* **2019**, *64*, 1239–1251. [CrossRef]
72. Esters, L.; Breivik, Ø.; Landwehr, S.; Ten Doeschate, A.; Sutherland, G.; Christensen, K.H.; Bidlot, J.R.; Ward, B. Turbulence scaling comparisons in the ocean surface boundary layer. *J. Geophys. Res. Oceans* **2018**, *123*, 2172–2191. [CrossRef]
73. Ellertsen, B.; Fossum, P.; Solemdal, P.; Sundby, S.; Tilseth, S. A case study of the distribution of cod larvae and availability of prey organisms in relation to physical processes in Lofoten. In Proceedings of the Propagation of Cod *Gadus morhua* L., Arendal, Noway, 14–17 June 1983.
74. Kristiansen, T.; Vollset, K.W.; Sundby, S.; Vikebø, F. Turbulence enhances feeding of larval cod at low prey densities. *ICES J. Mar. Sci.* **2014**, *71*, 2515–2529. [CrossRef]
75. Greenberg, D.A. Modelling the mean barotropic circulation in the bay of Fundy and Gulf of Maine. *J. Phys. Oceanogr.* **1983**, *13*, 886–904. [CrossRef]

© 2020 by the authors. Licensee MDPI, Basel, Switzerland. This article is an open access article distributed under the terms and conditions of the Creative Commons Attribution (CC BY) license (http://creativecommons.org/licenses/by/4.0/).

MDPI
St. Alban-Anlage 66
4052 Basel
Switzerland
Tel. +41 61 683 77 34
Fax +41 61 302 89 18
www.mdpi.com

MDPI Books Editorial Office
E-mail: books@mdpi.com
www.mdpi.com/books

Lightning Source UK Ltd.
Milton Keynes UK
UKHW052139120922
408735UK00002B/116